On Evolution

On Evolution

—⚏—

Charles Darwin and the Russian Prince, First
Nations and Twelve Step Societies

James G. Duncan

Foreword by Elizabeth May

© 2018 James G. Duncan
All rights reserved.

ISBN-13: 9781975990794
ISBN-10: 197599079X

To
my parents, my siblings,
and to my Higher Power

*Teach your children what we have taught our children —
that the Earth is our Mother.
Whatever befalls the Earth befalls the sons and daughters of the Earth.
This we know.
The Earth does not belong to us; we belong to the Earth.
This we know.
All things are connected — like the blood that unites one family.
All things are connected.
Whatever befalls the Earth befalls the sons and daughters of the Earth.
We did not weave the web of life
We are merely a strand in it.
Whatever we do to the web,
We do to ourselves...*

Chief Seattle

Table of Contents

Foreword ·xiii
Acknowledgements · xv
Preface · xvii
The Twelve Steps of Alcoholics Anonymous · · · · · · · · · · · ·xix
The Twelve Traditions of Alcoholics Anonymous · · · · · · · · · · · · ·xxi

Chapter 1 Change Ourselves and Change the World · · · · · · · · 1
    Chapter by Chapter · 8
Chapter 2 The Twelve Steps and the "gentle Russian prince" · · 11
    Defining the Twelve Step Movement · · · · · · · · · · · · · · · 12
    Picking up the thread of HPMA · · · · · · · · · · · · · · · · · 15
        Carl Jung · 15
        Old friends: Bill W. and Ebby T. · · · · · · · · · · · · · · 16
        Bill W.'s spiritual awakening to Dr. Bob · · · · · · · · · · 18
        "Flying Blind": The Oxford Group to A.A. · · · · · · · · 22
    The Twelve Traditions and the Russian prince · · · · · · · · · 26
    Who Was Peter Kropotkin? · 32
Chapter 3 Darwin and Kropotkin · **45**
    Cooperation, Mutual Aid, and Sociability · · · · · · · · · · · 48
    Darwin's Metaphor, and Contradiction · · · · · · · · · · · · 51
    Was Darwin racist? · 60
    Variability and violence in *The Origin of Species* · · · · · · · · · 64
    HPMA, Evolution, and the Future: Brief Notes · · · · · · · 66
Chapter 4 First Nations and Twelve Step Societies · · · · · · · · · **72**
    Authority in First Nations Societies · · · · · · · · · · · · · · · 77
    ". . . all virtue in a certain gentleness . . ." · · · · · · · · · · · 80
    Leadership in First Nations societies · · · · · · · · · · · · · · 86

    Fear and First Nations HPMA ························ 94
        Magellan then Drake in Tierra del Fuego ········ 95
        Darwin aboard the *Beagle* ···················· 97

Chapter 5 First Nations and Twelve Step HPMA ··········· 100
    Settings ····································· 100
    A Comparison of First Nations circles and Twelve
    Step groups ·································· 104
    The Sleepers of Safety ························· 105
        Allow all to speak without interruption ········ 106
        Do not comment on what others say ············ 106
        Keep the focus on yourself, not on others ········ 107
        Keep what you hear confidential ·············· 108
    Spiritual Parallels ····························· 109
    Children ···································· 111
    Ecology ····································· 114
    Other Considerations in Group Process ··········· 115

Chapter 6 Further along the Way Ahead ················ 121
    Chewing on Ideas ····························· 123
    The Individual in Relation ······················ 126
    Natural Ambiguity as Truth ···················· 130
        Discussion ······························· 133
    Humanly Knowable Truth ······················ 137
    Imperfection ································· 140
    Dialogue ···································· 142
    Citizenship ·································· 143
    What to Change? ····························· 145

Chapter 7 Twelfth Step work "in all our affairs" ··········· 147
    A.A. Twelfth Step work "in all our affairs" ········· 148
        Alcoholics Anonymous (1939) ················ 149
        Twelve Steps and Twelve Traditions (1953) ········ 151
        Alcoholics Anonymous Comes of Age (1957) ······ 154
        Other writings by Bill W. ···················· 155
    Al-Anon Twelfth Step work "in all our affairs" ······ 161
        Al-Anon literature ························ 162

"In all our affairs": Confusion and Uncertainty?······172
Personal Recovery·······178
Respect for myself·······181
Respect for the other·······182
Respect for what keeps us God's children········182
Serenity·······185
HPMAGs·······186
Chapter 8 Blue Green Politics ············193
Theory·······200
Practice·······207
Choices·······209
"Cradle-to-grave" product responsibility········212
Ecological moral investment·······212
Fair market pricing·······213
Alignment of accounting and ecological principles···213
HPMA and Democracy: One Step Ahead···········214
Chapter 9 Conclusions ·············217

Epilogue·······229
Appendix A: Kropotkin's Wider Influence·············235
Mutual Aid·······236
Self-Sacrifice for the Common Welfare···········239
Society·······245
Other Possible Influences from Kropotkin··········246
Appendix B: Darwin was not an ecologist············253
Appendix C: Nomad·······259
Appendix D: Reflections·······271
Appendix E: Permissions·······297
Bibliography·······299
Name Index·······333
Subject Index·······339
About the Author·······343

Foreword

JAMES DUNCAN HAS shared an intimate view of the Twelve Step program of Alcoholics Anonymous, founded by Bill Wilson and Dr. Bob Smith, June 10, 1935. Those with a passing familiarity will assume that A.A. is a program developed exclusively for those battling alcoholism. However, as the author points out, the A.A. program was actually conceived with additional, much wider goals in mind.

Duncan explores the deeper vision of Wilson and Smith in aspiring to change all aspects of an individual and of the wider society in which we live.

There is certainly a need to address addiction issues. Duncan refers to "money abuse." Modern industrialized society is awash in addictions. They distract us; consume us; anaesthetize us against the growing awareness that we are on a suicidal course.

We are addicted to fossil fuels. We are addicted to consumerism. We are told that buying more stuff will make us happy. Consumer confidence is the measurement of our economic well-being. Shopping translates to patriotic duty.

So, as in the A.A. program, the first step society needs to take is to admit, "…we [are] powerless over [our addictions] . . . that our lives [have] become unmanageable."

Beyond that, we need to seek answers. What truly makes us happy? How do we resolve our addictions? How do we place the common good above our private fears? How do we take control of our own lives?

Any wisdom that can be shed on these questions makes this book worth reading.

Elizabeth May
Federal Green Party leader, MP, OC

Acknowledgements

FIRST OF ALL, I must thank First Nations, Alcoholics Anonymous (A.A.), and the Al-Anon Family Groups (Al-Anon) for allowing me to explore my way through their spiritual ways of living. The many First Nations circles and Twelve Step groups that I attended over a 30 year period are where I found the idea and inspiration to write this book.

There are many people that I need to thank for their love and support, especially my parents, and my 13 siblings, Pat, Theresa, Betty, Donnie, Kathy, Margie, Ellen, MaryLou, Lorna, Laurie, Joey, Frances, and Christine. I must not forget the support of sisters-in-law and brothers-in-law.

Schools in Saskatchewan and Ontario where I received my early education deserve much credit, as does the University of Toronto where I learned to be a student. Many thanks are due to Phil Keddie, my MA thesis advisor at the University of Guelph, to Roland Lecomte, my MSW advisor at Carleton University in Ottawa, and to Agriculture Canada and Environment Canada where I spent more years of learning.

I also wish to thank Georges Sioui for providing me with the inspiration to write about Charles Darwin, and Elizabeth May for writing the foreword to this book. Lastly, my gratitude extends to the Create Space team at Amazon, for their kindness, patience, and professional work in helping me to publish this book.

Other people and institutions to whom I am grateful remain numerous and anonymous, in whole because of the failure of memory.

Preface

When I first titled this book *Getting Back to the Garden*, I was reminded by a First Nations woman that the Garden is still here, that we have not lost it, nor have we completely banished ourselves from it. What is left of its former glory sustains us still, after the ravages of the last 3 thousand years. Thus, we need not go back in time to find the Garden; indeed, that would be impossible. Rather, the Garden that the Earth could be again, and the path to it, is hidden, not in a specific geographical location, but in the recesses of our bodies, minds, and spirits, waiting there to unfold before us, to emerge and blossom once again.

In this book, I aim to carry the Twelve Step message "in all our affairs," to point to that path, an ancient, now revived, way of living that can bring us back to the Garden. Finding the path is not always easy, but walking along it is less difficult than it seems. Though many may tread the path with us, each person must decide to walk it, to find sobriety, sanity, serenity, happiness, and health, and to offer an example to those seeking a new, moral, spiritual approach to living in a world beset with profound illnesses.

I believe that the way ahead is in rejoining the spiritual avenue still thriving among ancient First Nations and now in the younger Twelve Step societies. Drawing from a range of sources, I try to fashion a coherent view of benign anarchic and democratic practices in community development, long ago forgotten or discarded by mainstream societies. With the understanding that evolution begins with individuals in First Nations circles and Twelve Step groups, I examine the theory and practice of their approaches to social architecture that can steer us to a broad highway leading to moral, scientifically sound, and peaceful global democracy befitting the New Age.

The Twelve Steps of Alcoholics Anonymous

1. We admitted we were powerless over alcohol - that our lives had become unmanageable.
2. Came to believe that a Power greater than ourselves could restore us to sanity.
3. Made a decision to turn our will and our lives over to the care of God *as we understood Him*.
4. Made a searching and fearless moral inventory of ourselves.
5. Admitted to God, to ourselves, and to another human being the exact nature of our wrongs.
6. Were entirely ready to have God remove all these defects of character.
7. Humbly asked Him to remove our shortcomings.
8. Made a list of all persons we had harmed, and became willing to make amends to them all.
9. Made direct amends to such people wherever possible, except when to do so would injure them or others.
10. Continued to take personal inventory and when we were wrong promptly admitted it.
11. Sought through prayer and meditation to improve our conscious contact with God *as we understood Him*, praying only for knowledge of His will for us and the power to carry that out.
12. Having had a spiritual awakening as the result of these steps, we tried to carry this message to alcoholics, and to practice these principles in all our affairs.

The Twelve Traditions of Alcoholics Anonymous

1. Our common welfare should come first; personal recovery depends upon A.A. unity.
2. For our group purpose there is but one ultimate authority - a loving God as He may express Himself in our group conscience. Our leaders are but trusted servants; they do not govern.
3. The only requirement for A.A. membership is a desire to stop drinking.
4. Each group should be autonomous except in matters affecting other groups or A.A. as a whole.
5. Each group has but one primary purpose - to carry its message to the alcoholic who still suffers.
6. An A.A. group ought never endorse, finance, or lend the A.A. name to any related facility or outside enterprise, lest problems of money, property, and prestige divert us from our primary purpose.
7. Every A.A. group ought to be fully self-supporting, declining outside contributions.
8. Alcoholics Anonymous should remain forever non-professional, but our service centers may employ special workers.
9. A.A., as such, ought never be organized; but we may create service boards or committees directly responsible to those they serve.
10. Alcoholics Anonymous has no opinion on outside issues; hence the A.A. name ought never be drawn into public controversy.

11. Our public relations policy is based on attraction rather than promotion; we need always maintain personal anonymity at the level of press, radio, and films.
12. Anonymity is the spiritual foundation of all our traditions, ever reminding us to place principles before personalities.

CHAPTER 1

Change Ourselves and Change the World

Never doubt that a small group of thoughtful, committed citizens can change the world. Indeed, it's the only thing that ever has.

MARGARET MEAD

WHY IS THERE such current enthusiasm about Alcoholics Anonymous ("A.A.") and the Twelve Step movement that followed its creation? At the turn of the millennium, we saw the publishing of two more biographies of William Griffith Wilson ("Bill W."), the visionary leader of A.A.; along with Dr. Robert Smith ("Dr. Bob"), he was A.A.'s cofounder. The new books were *Bill W.: A Biography of Alcoholics Anonymous Cofounder Bill Wilson* (Francis Hartigan, New York: St. Martin's Press, 2000), and *Bill W. and Mr. Wilson: The Legend and Life of A.A.'s Cofounder* (Matthew Raphael, Amherst, MA: University of Massachusetts Press, 2000). These add to the growing list of writers who have estimated Bill W. to be a man whose life and ideas are well worth writing about.

In the first decade of the 2000s, three more books about the Twelve Step movement made their appearance. *My Name is Bill* (2004), by Susan Cheever, serves up a detailed story of the man who transformed the way the globe deals with addiction.[1] Thomas Keating cogently examines spiritual aspects of the Twelve Steps in *Divine Therapy and Addiction* (2009).[2] And

---

1 Susan Cheever, *My Name is Bill*, (New York: Washington Square Press, 2004).
2 Thomas Keating, *Divine Therapy and Addiction, (New York: Lantern Books, 2009).*

James G. Duncan

Marya Hornbacher, in 2010, wrote a candid analysis of her own positive experience in Alcoholics Anonymous.³

Partial autobiographies by Bill W. were published in 1939 in *Alcoholics Anonymous* (the Big Book), and more recently by Hazelden in *Bill W.: My First 40 Years* (2000), from a 1954 audiotaping. After Bill W.'s death in 1971, a biography was written by Robert Thomsen in 1975 titled simply, *Bill W.* Too, Alcoholics Anonymous published its official biography of him in 1984, *'PASS IT ON': The story of Bill Wilson and how the A.A. message reached the world.*

To a degree, the Twelve Step movement has already changed the world. In his book *The Different Drum: Community Making and Peace*, Scott Peck described A.A. as the "most successful community . . . probably in the whole world,"⁴ and in a 1991 audiotaping said that A.A. was the "spearhead of the community movement." Hartigan concluded that a "community"⁵ of Twelve Step societies began with A.A. Its membership is estimated at 2 million people, and recovery from alcoholism takes place in over 100,000 groups in 150 countries worldwide.⁶ Say Doyle et al, 10 to 12 percent of Americans, though not alcoholic, are "alcohol dependent."⁷ This yields a shocking number of about 40 million Americans affected by dependency drinking, greater than the entire population of Canada.

The second Twelve Step society, the Al-Anon Family Groups ("Al-Anon"), for the families and friends of alcoholics, has grown to about

---

3 Marya Hornbacher, *Mental Illness, Addiction, and the 12* Steps, (Center City, MN: Hazelden, 2010).

4 M. Scott Peck, *The Different Drum: Community Making and Peace*, (New York: Simon & Schuster, 1987), 77.

5 Francis Hartigan, *Bill W.: A Biography of Alcoholics Anonymous Cofounder Bill Wilson*, (New York: St. Martin's Press, 2000), 212.

6 Alcoholics Anonymous, *Alcoholics Anonymous* (The Big Book), 4th edition, (New York: Alcoholics Anonymous World Services, Inc., 2001), xxiii.

7 Dr. Robert Doyle et al, *Almost Alcoholic: Is My (or My Loved One's) Drinking a Problem?* (Center City, MN: Hazelden Publishing, 2012), 30.

On Evolution

30,000 groups in over 112 countries[8] with a membership approaching 400,000. Al-Anon recognizes "the immense power for good" that is released by its "worldwide network of groups."[9] Notably, Anne Lawson et al state that "Research . . . has suggested that the best treatment predictor of an alcoholic's sobriety is his or her involvement in Alcoholics Anonymous . . . coupled with the spouse's involvement in Al-Anon."[10]

Matthew Raphael thought that Bill W. wanted alcoholics to have only an "inside job": a quasi-religious conversion experience,"[11] and contends further that Bill W. did not treat excessive drinking as a "structural or social problem." Yet, as proof, Raphael cites only "Bill's Story" in the Big Book. "Bill's Story" is about the cofounder's drinking years and early recovery, not his difficult childhood in an extended family marked by too much drinking. Nor is the story about the decades-long growth and maturing of A.A. under the watchful eyes of the cofounders, Bill W. and Dr. Bob.

Charles Bufe wrote about the issue of social change in A.A., also concluding that A.A. was not interested in changing the world.[12] Klaus Makela and his international team of social scientists describe A.A. as a "mutual-help movement" in community development, but in spite of his

---

8 *The Al-Anon Family Groups, Classic Edition*, 1955 original edition with footnotes and annotations added in 2000, (Virginia Beach, VA: Al-Anon Family Groups Headquarters, Inc., 2000), 6.

9 *One Day at a Time in Al-Anon*, (Virginia Beach, VA: AFG Headquarters, Inc., 1973), 201. Al-Anon has not yet reached its potential, not nearly, and may assume more of a leadership role in combating alcoholism. Bill W. felt that Al-Anon would become larger than A.A. "because each alcoholic affected about five other people." (*First Steps...35 Years of Beginnings*, [Virginia Beach, VA: AFG Headquarters], 1986, 61).

10 Anne Lawson et al, *Alcoholism and the Family: A Guide to Treatment and Prevention*, 2nd edition, (Gaithersberg, MD: Aspen Publishers, Inc., 1998), 316.

11 Matthew Raphael, *Bill W. and Mr. Wilson: The Legend and Life of A.A.'s Cofounder*, (Amherst, MA: University of Massachusetts Press, 2000), 55.

12 Charles Bufe, *Alcoholics Anonymous: Cult or Cure?* (San Francisco: Sharp Press, 1991), 101.

observation that 260 separate Twelve Step fellowships[13] were in existence by 1991,[14] he does not describe the potential of the Twelve Step movement for wide and deep social change, locally and globaly.

Quite opposite to the views of Raphael, Bufe, and Makela, the Twelve Step movement is *all about* changing the world - by first changing the self. The movement is about living the Twelve Steps to change ourselves and change the world, by discovering ourselves in relation to the other, and by learning the moral, spiritual principles of a better way of living. Twelve Step societies promote individual healing within the common welfare, while the message is carried to others and the suggested principles[15] are practiced by individuals and groups in all their affairs.[16] Thus, social change is inherent in the individual change that takes place in the near-autonomous Twelve Step groups.

We can approach this subject in an organized fashion. In so doing, we will see that the Twelve Step movement is not only social, but ecological, with the potential to move the world, the lever about which Archimedes theorized on the physical plane.

While changing the self is the first priority of Twelve Step groups, there is less awareness that the pioneers of its first two societies, A.A. and Al-Anon, wanted to change the world, too, to make it a better and safer place for all. A.A.'s biography of Bill W. says that he "believed [A.A.] would change the world."[17] Bill W.'s wife, Lois W., cofounder of the Al-Anon

---

13 Though "fellowship" was first used to describe A.A., "society" was employed later as well. Bill W. used the two words almost interchangeably in *The Language of the Heart: Bill W's Grapevine Writings*, (New York: The A.A. Grapevine, Inc., 1988).

14 Klaus Makela et al, *Alcoholics Anonymous as a Mutual-Help Movement: A Study in Eight Societies*, (Madison, WI: University of Wisconsin Press, 1996), 216. William White provides a 1998 estimate of over 400 Twelve Step societies in *Slaying the Dragon: The History of Addiction Treatment and Recovery in America*, (Bloomington, IL: Chestnut Health Systems/ Lighthouse Institute, 1998), 163.

15 The principles of Twelve Step practice are suggested only; no form of coercion is involved. This book can be considered one long suggestion.

16 See A.A.'s Step Twelve at the beginning of the book.

17 Alcoholics Anonymous, '*PASS IT ON*': *The story of Bill Wilson and how the A.A. message reached the world*, (A.A. World Services, Inc., 1984), 265.

On Evolution

Family Groups, said it was no exaggeration that her husband wanted to transform the world for the better.[18] She also stated her own beliefs in the potential of A.A., Al-Anon, and Alateen for broad beneficial change in the Al-Anon book, *First Steps: Al-Anon...*[19] and in the Al-Anon audiotapes, *Lois and the Pioneers*. Bill W., Dr. Bob and his wife Anne S., foresaw the world-changing power of their emerging program as early as 1937,[20] before their group was named Alcoholics Anonymous in 1938, and before *Alcoholics Anonymous* (the Big Book) was published in 1939.

The Twelve Step pioneers were not alone in their estimation of the potential of the movement to combat alcoholism, along with drug abuse, the foremost plague of the planet.[21] Bill W.'s doctor, William Silkworth, began in 1939 with his assessment of the great good that came from A.A.[22] The Jack Alexander article in the *Saturday Evening Post* propelled A.A. into the limelight on March 1, 1941.[23] By 1946, R.G. McCarthy stated that A.A.'s success was "perhaps without parallel in our society."[24] Norman Vincent Peale, in about 1950, thought that A.A. was the greatest spiritual

---

18 Al-Anon Family Groups, *First Steps: Al-Anon . . . 35 Years of Beginnings*, (Virginia Beach, VA: AFG Headquarters, Inc., 1986), 156.

19 Ibid

20 *Alcoholics Anonymous Comes of Age*, (New York: A.A. World Services, Inc., 1957), 76.

21 Browne-Miller says that there are 2 *billion* alcohol users in the world, outdistancing smokers and drug users combined. (Angela Browne-Miller, *The Praeger International Collection on Addictions. Volume 1: Faces of Addiction, Then and Now*, [Westport, CT: Praeger, 2009], xi). She adds that, "[The] scourge of addiction to alcohol and other drugs is one of the most serious and urgent issues facing humanity today... [it] is epidemic in scale...if we fail to fight [it] with all available means, [it] will soon be out of control worldwide..." (xxv-xxvi).

An article in the *Ottawa Citizen* (December 20, 2014, I1) said that excessive alcohol consumption may have "replaced tobacco as the number one preventable health scourge in Canada." It noted too that heavy drinking costs $14 billion in health care, law enforcement, lost productivity...while bringing in only $9 billion in federal and provincial taxes, licensing fees, etc.

22 *Alcoholics Anonymous* (The Big Book), (New York: A.A. World Services, Inc.), xxiii–xxx.

23 *Alcoholics Anonymous Comes of Age*, (New York: A.A. World Services, Inc.), 16.

24 David Robinson, *Talking Out of Alcoholism: The Self-Help Process of Alcoholics Anonymous*, (London: Crown Helm, 1978), 9.

force in the world.[25] O.H. Mowrer wrote in 1961 that *only* A.A. had shown the world a promising spiritual path for the future.[26] Pittman et al noted the striking effectiveness of A.A. in *Society, Culture, and Drinking Patterns* (1962).[27] Analyst Marion Woodman said in her 1982 book, *Addiction to Perfection*, that she encourages her clients to go to A.A. meetings.[28] By 1987, Scott Peck felt that A.A. was "without doubt the single most effective agency of human transformation in our society."[29] Nan Robertson told the tale, in 1988, of the definite effectiveness of her own A.A. program.[30] Anne Wilson Schaef gave much praise to Twelve Step programs in 1987, 1988, and 1992, saying in 1992 that they are the best tools for recovery from addictions;[31] she sends her trainees to A.A. meetings. "Experience shows," said Hirschfield in 1990, "that [the Twelve Step program] works for *all* those who want it."[32] Dupont et al (1994) described the Twelve Step movement as a ". . . quiet and powerful revolution," capable of being ". . . the best chance of winning the war against alcohol and drug abuse, and of rebuilding stronger, healthier families and communities."[33] In 1996, Makela et al called A.A. "one of the great success stories of the [20th]

---

25 Al-Anon Family Groups, *Lois Remembers*, (Virginia Beach, VA: AFG Headquarters Inc., 1979), 143.

26 O.H. Mowrer, *The Crisis in Psychiatry and Religion*, (Princeton, NJ: Van Nostrand, 1961), 109.

27 David Pittman et al, eds., *Society, Culture, and Drinking Patterns*, (New York: John Wiley and Sons, Inc., 1962), 548.

28 Marion Woodman, *Addiction to Perfection*, (Toronto: Inner City Books, 1982), 28.

29 M. Scott Peck, *The Different Drum*, 185.

30 Nan Robertson, *Getting Better: Inside Alcoholics Anonymous*, (New York: William Morrow and Company, Inc., 1988).

31 Anne Wilson Schaef, *Beyond Therapy, Beyond Science: A New Model for Healing the Whole Person*, (San Francisco: Harper and Row, 1992), 129, 131, 282.

32 Jerry Hirschfield, *The Twelve Steps for Everyone . . . who really wants them*, revised edition (Center City, MN: Hazelden Publishing, 1990), 15.

33 Robert Dupont et al, *A Bridge to Recovery: An Introduction to Twelve-Step Programs*, (Washington, DC: American Psychiatric Press, Inc., 1994), xviii.

century."[34] Breton et al said, also in 1996, that society is addicted, and use the Twelve Step model to advance their ideas about individual and societal recovery.[35] Deepak Chopra said in 1997 that Alcoholics Anonymous is "one of the most powerful approaches to alcoholism ever developed," largely due to its spirituality.[36] Room[37] and Valverde[38] share much optimism for the healing potential of the Twelve Step movement. Room, in addition, has said that A.A. is now an international social movement.[39] In 2001, Fuller stated that A.A.'s "open-minded and eclectic approach to spiritual regeneration makes it one of the most powerful mediators of wholeness . . . today."[40] Also in 2001, Mitrof wrote - I believe with no exaggeration - that A.A.'s [program is] among ". . . the most important ideas ever invented by humans."[41] Simos cites A.A.'s strength in her 2004 book, *The Earth Path*[42] and Heyman said simply in 2009, "AA works."[43] McGovern

---

[34] Klaus Makela et al, *Alcoholics Anonymous as a Mutual-Help Movement*, 3.

[35] Denise Breton et al, *The Paradigm Conspiracy: Why Our Social Systems Violate Human Potential - And How We Can Change Them*, (Center City, MN: Hazelden, 1996).

[36] Deepak Chopra, *Overcoming Addictions: The Spiritual Solution*, (New York: Harmony Books, 1997), 51-52.

[37] Robin Room, "Healing Ourselves and Our Planet: The Emergence and Nature of a Generalized Twelve-Step Consciousness." (*Contemporary Drug Problems* 19, no. 4 (1992): 717–40); Room, "Alcoholics Anonymous as a Social Movement" (Barbara McCrady et al, eds., *Research on Alcoholics Anonymous: Opportunities and Alternatives*, [New Brunswick, NJ: Rutgers Center of Alcohol Studies, 1993], 167–87).

[38] Mariane Valverde et al, "One Day at a Time and . . ." (*Sociology: The Journal of the British Sociological Association*, 33, no. 2 (1999): 393-410).

[39] The numerous other Twelve Step societies, including the Al-Anon Family Groups, Narcotics Anonymous, and Nicotine Anonymous, are also a part of this movement.

[40] Robert Fuller, *Spiritual but Not Religious: Understanding Unchurched America*, (New York: Oxford University Press, 2001), 115.

[41] William Hammond III, ed., *12 Step Wisdom at Work: Transforming Your Life and Your Organization*, (Dover, NH: Kogan Page U.S., 2001), vii.

[42] Miriam Simos, *The Earth Path: Grounding Your Spirit in the Rhythms of Nature*, (San Francisco: Harper Collins Publishers, 2004), 207.

[43] Gene Heyman, *Addiction: A Disorder of Choice*, (Cambridge, MA: Harvard University Press, 2009), 20.

said the same thing in Hornbacher's 2010 volume, *Sane,* that . . . peer recovery groups such as Alcoholics Anonymous . . . really work."[44] In 2015, Joseph Nowinski gave us a scientific view that supports the efficacy of A.A.[45] Nancy Ellen Abrams took this approach, also in 2015, saying that the Twelve Step movement could bring a planetary morality to a humanity awaiting inspiration.[46] Russell Brand, speaking from experience, gave his wholehearted support for the Twelve Step movement in 2017.[47] And Glenn Chesnut said beautifully, also in 2017, "[W]e are seeing the development, before our very eyes, of a vision of the world that works. How could the good God...not be delighted by the countless flowers of grace which AA has sprinkled all about the earth."[48]

Alcoholics Anonymous, the Al-Anon Family Groups, and the many other Twelve Step organizations, are an almost miraculous international movement for individual and social, ecological transformation. Enthusiasm for A.A. and its followers is well founded. We can trace its growth as a living thread, now become a durable fabric that many can wear for the healing of many ills, as foreseen by the Twelve Step pioneers. It is instructive to follow this thread, to allow us to see how Twelve Step practice promotes morality and a better way of living for individual and social well-being.

Chapter by Chapter

The second chapter of the book follows the evolution of Alcoholics Anonymous from 1930 through to the mid-1940s, with a cast of

---

44 Marya Hornbacher, *Sane: Mental Health, Addiction, and the Twelve Steps,* (Center City, MI: Hazelden Publishing, 2010), v.

45 Joseph Nowinski, *If You Work It, It Works! The Science Behind 12 Step Recovery,* (Center City, MN: Hazelden Publishing, 2015). Nowinski and others do what A.A., sadly, cannot: promote its program. See A.A.'s 11th Tradition at the front of the book.

46 Nancy Abrams, *A God That Could Be Real: Spirituality, Science, and the Future of Our Planet,* (Boston: Beacon Press, 2015).

47 Russell Brand, *Recovery: Freedom from Our Addictions,* (London: Pan Macmillan, 2017).

48 Glenn F. Chesnut, *Father Ralph Pfau and the Golden Books: The Path to Recovery from Alcoholism and Drug Addiction,* (Bloomington, IN: iUniverse, 2017), 193.

On Evolution

characters including Bill and Lois W.; Rowland H.; Ebby T.; Dr. Bob and his wife, Anne S.; William James; Carl Jung; and the "gentle Russian prince" of *Alcoholics Anonymous Comes of Age,* Peter Kropotkin. I introduce here a description of A.A. and Al-Anon as "Higher Powered Mutual Aid," borrowing higher power from James and mutual aid from Kropotkin. People gather for mutual aid in First Nations circles and Twelve Step groups, guided by Higher Powers; thus, these circles and groups practice Higher Powered Mutual Aid, or HPMA.

Chapter Two includes the similar ideas of Bill W. and Kropotkin on the subject of mutual aid that A.A. and Al-Anon groups, and First Nations circles, practice so well. Their thinking is contrasted with that of Charles Darwin who preached only competition, not cooperation, sociability, or mutual aid, in his "Struggle for Existence."[49]

In Chapter Three, we compare and contrast some of the evolutionary ideas of Darwin and Kropotkin, important as they are for the further development of HPMA in the world. I observe and maintain that Darwin's competitive theory of evolution has no place in First Nations or Twelve Step cultures, while Kropotkin's opposing cooperative theory of evolution definitely does. This has important implications.

I point to a way ahead in the fourth and fifth chapters, describing significant parallels between age-old First Nations HPMA societies and the more recent Twelve Step. The suggestion is that lessons learned and exchanged between them can help us change ourselves and change the world. Included here are more important words regarding the evolutionary outlooks of Darwin and Kropotkin.

In these latter chapters especially, I have used informal participant observation, or field research, first set out by Lofland (1971), and expanded upon by, among others, Patton (1990) and Neuman (1994). It is mainly in hindsight, through the recollection of my own recovery in Twelve Step groups, A.A. and Al-Anon, and First Nations circles - the

---

49 The term "Struggle for Existence" was first defined by Darwin in *The Origin of Species*, (Charles Darwin, *The Origin of Species*, 1st edition 1859, [Harmondsworth, UK: Penguin Books, 1968], 116).

Pinganodin ("Silent Wind") Men's Lodge, sweat lodge and medicine wheel ceremonies, Kumik ("Elders Lodge"), and "Circle of Nations" - that I have gathered together the observations I share with the reader.

"First Nations," rather than Indigenous or Aboriginal, is used here because I focus on what I know. I have little knowledge of Inuit or Métis culture, and almost none of other Indigenous cultures. "Indian" is not used because I believe it should apply only to the people of India; it is past time to do so.

Chapter Six, "Further Along the Way Ahead," explores the vital concepts of the "individual in relation" and "humanly knowable truth," both central to human purpose and to the full practice of HPMA. The exercise of Twelve Step principles "in all our affairs"[50] is examined in the Seventh Chapter, including in the democratic and ecological spheres. Using my own knowledge gained in life, education, and recovery, I make the case that such practice carries the potential for far greater good than is recognized in the world today.

The Eighth Chapter, "Blue Green Politics," outlines the promise of HPMA for worldwide democratic, ecological, and spiritual renewal. I introduce the theory and practice of "Blue Green Politics" as a means of seeing ourselves and the world in a holistic light, and of changing our politics to match the urgent needs of the third millennium.

The Ninth and final chapter summarizes our findings, and presents tentative conclusions about using Higher Powered Mutual Aid, the First Nations and Twelve Step varieties, to change ourselves and change the world. To accomplish this, the chapter will further conclude with a "Blue Green" attempt to bring a deeply sleeping humanity to its much needed wakening.

---

50 See A.A.'s Twelfth Step at the front of the book.

CHAPTER 2

The Twelve Steps and the "gentle Russian prince"

If you give me an egg and I give you an egg,
we each have one egg.
If you give me an idea and I give you an idea,
we each have two ideas.

WEST AFRICAN PROVERB

IN THIS CHAPTER, we examine certain aspects of Twelve Step HPMA, where it came from, and some of the basics of its evolution. Behind the stories of Bill W.'s stunning spiritual awakening and of the supporting pillars of medicine and religion, the history of Alcoholics Anonymous is the story of the courageous struggle of men and women to choose life over death - using hard work, mutual aid, and the guidance of Higher Powers - in the midst of the Great Depression of the 1930s.

Less obvious are key connections with William James, Carl Jung, and Peter Kropotkin. "Outside issues" are studiously avoided in A.A., Al-Anon, and other Twelve Step societies,[51] and works by others are rarely cited. Nevertheless, the ideas of James, Jung, and Kropotkin greatly strengthened the foundations of A.A. and the basics of the Twelve Step movement, tightening its hold in the Earth itself.

---

51 Though "fellowship" was first used to describe A.A., "society" was employed later as well. Bill W. used the two words almost interchangeably in *The Language of the Heart: Bill W's Grapevine Writings*, (New York: The A.A. Grapevine, Inc., 1988).

Defining the Twelve Step Movement

Twelve Step programs have been variously described as self-help, mutual help, mutual support, mutual aid, and perhaps in other ways. This proliferation of terms to describe the same thing seems unfortunately due to misperceptions and misunderstandings of Twelve Step programs, and their theoretical and practical foundations. Some of these terms might therefore be eliminated.

A.A. has only rarely described itself as "self-help" since its inception, and it is difficult to understand how it has come to be used so widely to describe Twelve Step HPMA. Katz et al (1976, 1993), Powell (1987), Wasserman (1988), Romeder (1990), Van Den Bergh (1991), Kaminer (1992), Lavoie (1994), Borkman (1999), Humphreys (2004), Alexander (2008), and Nowinski (2015) are among those who use "self-help" to define Twelve Step societies.[52] Self-help may have originated with professionals who, on observing that their help was not needed, might have said, "Oh, they are helping them*selves*." But self-help implies that the aim of Twelve Step HPMA programs is to assist people to help only their individual selves, a kind of constant taking but never giving.

Yet it has always been the case that people in Twelve Step groups and societies can only help themselves *by helping others* as well. Members are encouraged to focus on their own lives in order to become well, but one can only receive the benefits of participation in Twelve Step programs by engaging in the mutually beneficial process of receiving for oneself and giving back to the members of one's group. Twelve Step groups exist and thrive only if members give back to the group; the welfare of the group is essential for individual recovery and healing to take place. Thus, the term "self-help" is inappropriate when used to describe the Twelve Step movement and its HPMA programs and groups. Makela et al feel the

---

52 Some differentiate between "mutual aid" and "self-help" but go on to use them interchangeably. e.g., Alfred Katz et al, 1976; Francine Lavoie, 1994; T. Borkman, 1999.

On Evolution

same,[53] and William White, using "mutual aid" frequently, writes that, "...A.A. is the antithesis of self-help..."[54]

"Mutual support" seems timid, appearing to describe the propping up of the unfortunate ill who engage only in the sharing of each other's miseries; it is sometimes used by Twelve Step societies. However, the word "support" does not capture the dynamic process of recovery inherent in their groups. "Mutual help" is preferred by Makela et al, but, partly for historical reasons described later, "mutual aid" is a superior term when used to describe Twelve Step groups and First Nations circles.

It is essential to note that self-help, mutual support, mutual help, and mutual aid all fail to account for the central place of Higher Powers in First Nations and Twelve Step societies. Mutual aid theorists such as Katz et al (1976), Sullivan (1980), Barclay (1990), and Makela et al (1996), work in the right direction re: the great potential of mutual aid, but underestimate, or take for granted, the import of Higher Powers. Makela et al say that A.A. groups are "self-governing, subject to no external authority or superstructure,"[55] seeming to ignore the explicit recognition of Higher Authority in A.A.'s Tradition Two.[56]

Some have said that A.A. is uniquely American in its approach,[57] even that Bill W. and the founders had built an "American institution."[58] The Twelve Step movement had its origins in the United States but its heartland now includes Canada, with worldwide appeal. Bill W. emphasized that the Twelve Step principles are universal, borrowed from ancient sources,

---

53 Klaus Makela et al, *Alcoholics Anonymous as a Mutual-Help Movement*, 13.

54 William White, *Slaying the Dragon: The History of Addiction, Treatment and Recovery in America*. (Bloomington, IL: Chestnut Health Systems/Lighthouse Institute, 1998), 144.

55 Klaus Makela et al, 43.

56 See A.A.'s Twelve Traditions at the front of the book.

57 Ernest Kurtz, *Not God: A History of Alcoholics Anonymous*, (Center City, MN: Hazelden Educational Services, 1979), 182; Klaus Makela et al, 27.

58 Matthew Raphael, *Bill W. and Mr. Wilson*, 111.

and not unique, as some think.[59] Lois W. echoed her husband's feelings, saying that A.A. principles coincide with universal laws.[60]

While the Twelve Step pioneers made use of these laws, they did so in a creative, ingenious way that *was* unique, leaving them with the belief that the Twelve Step movement would make positive changes in the world. Led by Bill W., Dr. Bob, and their wives, Lois W. and Anne S., the founders transformed simple mutual aid into Higher Powered Mutual Aid, a form of social organization that enlists benevolent Higher Powers to help its cause. Higher Powered Mutual Aid is what ancient First Nations societies and newer Twelve Step fellowships do and are known for.

Use of the spiritual term Higher Power together with Mutual Aid would not have appealed to scientists like Russian prince Kropotkin, the man who first extensively used the phrase mutual aid in the late nineteenth and early twentieth centuries. Yet the combination of the two is an extremely potent one, and greater awareness and implementation of Higher Powered Mutual Aid in the world today would, I am convinced, help to ensure the future success of our species.

Other ecological, philosophical, psychological, moral, religious, and/or spiritual approaches to healing, and to the liberating of peoples' creative potential within the common welfare, could use HPMA principles and methods for beneficial change in individuals, families, communities, nations, and for the Earth as a whole. Higher Powered Mutual Aid, exemplified in the recent successes of the Twelve Step movement, and in far older First Nations societies,[61] has much unrealized potential for healing the wide rifts that divide and threaten to annihilate us today.

---

59 Alcoholics Anonymous, *Alcoholics Anonymous Comes of Age*, (New York: A.A. World Services, Inc., 1957), 224, 231; *The Language of the Heart*, 345.

60 *Alcoholics Anonymous Comes of Age*, 228.

61 Alfred Katz et al write of mutual aid among European settlers in the Americas, but say nothing of that practiced by the "Indians" the Europeans encountered along the way. (Alfred Katz et al, 1976, 18-19).

On Evolution

Picking up the thread of HPMA

How did Bill W. and the pioneers discover Higher Power Mutual Aid, or rather uncover its ancient roots? What was its value to Bill W. personally, the prime mover of A.A. and the Twelve Step movement? How did he and the pioneers put the HPMA principles into practice as effectively as they did? If we are to understand and build upon their achievements, it is important to answer such questions in an adequate manner.

After a difficult childhood in a home and extended family where alcohol was abused, Bill W. became a competitive lone wolf,[62] and an alcoholic, striving to be the "Number One Man" ahead of all others, to keep at bay the wolves of his own inadequacies. This is clear in *Alcoholics Anonymous* (1939),[63] in *Alcoholics Anonymous Comes of Age* (1957), in the autobiography of Bill W.'s wife, *Lois Remembers* (1979), in A.A.'s biography of Bill W., *'PASS IT ON': The story of Bill Wilson and how the A.A. message reached the world* (1984), and in *Bill W.: My First 40 Years* (2000).

The living fibre of Higher Powered Mutual Aid is a fuzzy one, with many small additional rootlets in the soils of the northeastern United States. It led to the formation of Alcoholics Anonymous in 1935, and to the second Twelve Step society, the Al-Anon Family Groups in 1951; the latter makes explicit use of the term "mutual aid" in its Third Tradition. There are almost as many places to start here as there are authors, but Bill W. gives credit for A.A.'s founding to meetings between Carl Jung and another American alcoholic, Rowland H., beginning in about 1931.

Carl Jung

The A.A. Grapevine article, "The Bill W. - Carl Jung Letters,'" is an exchange of correspondence between Bill W. and famous psychoanalyst Carl Jung,[64] recorded also in more detail in *'PASS IT ON'*.[65] Bill W. extended many thanks

---

62 Alcoholics Anonymous, *Alcoholics Anonymous* (The Big Book), 3.
63 Ibid, 1-16.
64 *The Language of the Heart*, 276-81.
65 Alcoholics Anonymous, *'PASS IT ON': The story of Bill Wilson*, 381–86.

to Jung for the help that A.A. had received from his meetings with Rowland H. He recounted that it was Jung's "humility" and "deep perception" in the Rowland H. case, which set the early stage for Alcoholics Anonymous.[66] From the same letter, Bill W. outlined his understanding of what Jung had confided to Rowland H.:

- That Jung was powerless to help him further.
- That Rowland H. should subject himself to a spiritual or religious atmosphere and pray for some sort of conversion experience or transformation.[67]

Jung's response to Bill W.'s letter confirmed his understanding of what had transpired between Jung and his client. However, the significance of their meetings would not come to the attention of the Twelve Step pioneers for some time. At about the time of the Jung-Rowland H. sessions, Bill W. was drinking heavily, and only on December 11, 1934, was he able to stop. A more complete description follows of the link between Jung's important insights and the development of Twelve Step HPMA.

OLD FRIENDS: BILL W. AND EBBY T.

In the early 1930s, Bill W. had been having a terrible time, drinking and in and out of hospital, with his wife Lois working to support them. An old friend, Ebby T., under the influence of the evangelical Christian Oxford Group[68] rather than alcohol, came to Bill W.'s aid in November 1934. Their long talk, and subsequent talks, together and with others, reveals

---

66 Ibid, 382–83.
67 Ibid, 382.
68 The Oxford Group was founded by Lutheran minister Frank Buchman in the early 1920s. (Alcoholics Anonymous, *Dr. Bob and the Good Oldtimers: A biography, with recollections of early A.A. in the Midwest*, 54-55); *'PASS IT ON': The story of Bill Wilson*, 130. The Oxford Group was most popular in the 1930s but lost influence in the 1940s. (William White, *Slaying the Dragon*, 128).

On Evolution

early application of the principles of HPMA, recorded in similar ways by Bill W.,[69] and by A.A.,[70] summarized here in point form:

- Rowland H. joined the Oxford Group and found sobriety, as Jung thought he might.
- In early 1934, Ebby T. joined the Oxford Group, persuaded by three friends, among them Rowland H., who had found Ebby in jail for alcohol related transgressions.
- Rowland H. influenced Ebby T. the most.[71] Using beautiful imagery, Bill W. described Ebby's conversion as the cell-like growth of the embryo of A.A.,[72] with three alcoholics talking to a fourth, Ebby T. This took place prior to Ebby's meetings with Bill W.
- Bill W.'s first meeting with Ebby was critical. After his friend refused a drink, Bill W. asked him, "...what has gotten into you? What is this all about?" Ebby T. replied, "I've got religion," and Bill wrote he "might as well have hit [me] in the face with a wet mop."[73]
- Saying that, "in no waking hour was the thought of my friend absent from my mind,"[74] Bill W. reflected on the realization that talking to another alcoholic was different than talking to professionals, such as his doctor, William Silkworth.[75] This radical idea proved to be the mainspring for the foundational development of HPMA in the Twelve Step movement.

---

69 *Alcoholics Anonymous* (The Big Book), 9–16; *Alcoholics Anonymous Comes of Age*, 58–77; *Bill W.: My First 40 Years*, (Center City, MN: Hazelden, 2000), 123–60.
70 *'PASS IT ON': The story of Bill Wilson*, 111–50.
71 Ibid, 114-15.
72 *Bill W.: My First 40 Years*, 130.
73 *'PASS IT ON': The story of Bill Wilson*, 111-12.
74 *Alcoholics Anonymous Comes of Age*, 59.
75 Ibid, 64. Dr. Silkworth assisted Bill W. enormously in the early 1930s.

- Acknowledging Bill W.'s anti-religious feelings, Ebby T. said that the Oxford Group was more spiritual than religious.[76] Responding to Bill W.'s difficulty in accepting a Higher Power, Ebby suggested too that he use his own conception of God.[77] He got these ideas from the Oxford Group, and Bill W. said it struck him intensely: he realized that he had only to believe in a Power greater than himself.[78]
- Bill W. wrote that Ebby T. said it was absolutely necessary to follow these Oxford Group principles in all aspects of life, and that it was essential to help others as he had helped me.[79]
- Ebby T. had made a "terrific impression" on Bill W.,[80] "at great depth," making him "ready for the gift of release," with "the missing link: one alcoholic talking to another, bearing hopelessness in one hand and hope in the other."[81]

Raphael writes, aptly, that Bill W. had received "a Twelfth Step call from his boyhood friend."[82] The description of how a fellow alcoholic had helped him - and how this was different from the professional medical help he had received - was Bill W.'s first awareness of how alcoholics could help one another. This was verified in his 1935 springtime meeting with Dr. Bob, when the embryo of A.A. issued forth as an infant.

Bill W.'s spiritual awakening to Dr. Bob

Drunk again after his meeting with Ebby T., Bill W. entered hospital for the fourth time, and his "hot flash" (as he put it later) spiritual awakening,

---

76 *Bill W.: My First 40 Years*, 113.

77 *Alcoholics Anonymous* (The Big Book), 12.

78 Bill W. said, "I had always believed in a Power greater than myself…I was not an atheist." (*Alcoholics Anonymous* (The Big Book), 10).

79 *Alcoholics Anonymous* (The Big Book), 14.

80 *Bill W.: My First 40 Years*, 149.

81 Ibid, 153–54.

82 Matthew Raphael, *Bill W. and Mr. Wilson*, 73.

On Evolution

was an amazing result. He thought the experience might be hallucination, but his doctor did not agree.[83] Lois W., always by his side, noticed a difference in his eyes,[84] the windows, it is said, to the soul. There has been much conjecture on the subject, recently by Hartigan[85] and Raphael.[86]

Bill W. did not drink again, read psychologist William James's *The Varieties of Religious Experience*[87] thoroughly, and began attending Oxford Group meetings in New York. He fervently pursued other alcoholics "morning, noon, and night"[88] and took up the study of alcoholism and its history, learning a lot from Dr. Silkworth. Gradually he found that he had to stop preaching to other alcoholics.

He tried working again, and on a bad day for business in Akron, Ohio in May 1935, he was sorely tempted to drink at a bar in the Mayflower Hotel. But, as Bill W. recalled later, he knew that he needed another alcoholic to talk to,[89] and the realization took him, most fortunately for Bill W., and for millions, to a nearby telephone and directory.

Many calls were placed by Bill W., first to Reverend Walter Tunks, and finally to Henrietta Seiberling, a loyal friend of Dr. Bob and his wife, Anne S., all Oxford Group members in Akron. Henrietta invited Bill W. over right away, seeing it as a great opportunity to help Dr. Bob stop his drinking. The next day's meeting between Bill W. and Dr. Bob, Mother's Day, May 12, 1935, was arranged by Anne S. and Henrietta Seiberling in

---

83 *Bill W.: My First 40 Years*, 148; *Alcoholics Anonymous Comes of Age*, 62; *'PASS IT ON': The story of Bill Wilson*, 123. The spiritual awakening is an epiphany of sorts for A.A. and other Twelve Step members, who can come to a *moral* awakening as well; 7 of the Twelve Steps deal with morality.

84 Al-Anon Family Groups, *Lois Remembers*, (Virginia, VA: AFG Headquarters, Inc.), 89.

85 Francis Hartigan, 60–62.

86 Matthew Raphael, 81–96.

87 William James, *The Varieties of Religious Experience*, (New York: Mentor Books, 1958, Originally published 1902). As noted, *Varieties* was very influential in the Oxford Group, and later in A.A.

88 *'PASS IT ON': The story of Bill Wilson*, 31.

89 Ibid, 136.

the latter's home. It confirmed what Bill W. had discovered during his talks with Ebby: that one alcoholic could help another to stay sober.

Perhaps wanting to see himself as the "Number One Man," Bill W. seems light on credit to the earlier conversation with Ebby T., saying that his talk with Dr. Bob was now the "final missing link."[90] The idea of one alcoholic helping another had also been "clinched"[91] in his talk with Ebby, who had also provided "the missing link."[92] Before going to the telephone in the Mayflower lobby, prior to meeting Dr. Bob, he recollected that he needed another alcoholic to talk to, to prevent himself from drinking.

Yet Bill W. recognized that the talk with Dr. Bob had a special importance - its mutual aid character.[93] Whereas Ebby T. had come to his friend's assistance, Bill W. was drinking during the conversation with Ebby and in no condition to help anyone else. Though he had been looking forward to meeting and drinking with his old pal, Ebby, Bill W. also said that he was happy to have all the gin to himself.[94]

Seen in this light, the conversation between Ebby T. and Bill W. was not as mutual as that of Bill W. and Dr. Bob. In the second conversation, neither of the men was drinking and each was seeking help from the other. More precisely, Bill W. was looking for help to stop himself from drinking, while Anne S. and Henrietta Seiberling were seeking assistance for Dr. Bob, and had to persuade him to attend the meeting.

Overlooked in most accounts of the meetings between Bill W. and Ebby T., and Bill W. and Dr. Bob, is the central element of the Higher Power. In Bill W.'s 1934 conversations with Ebby, only Ebby could speak about spiritual conversion. Before that talk, and before his own spiritual awakening, Bill W. had ambivalent feelings about God and religion. However, by 1935, Bill W., Dr. Bob, and their

---

90 *Alcoholics Anonymous Comes of Age*, 70.

91 Ibid, 64.

92 *Bill W.: My First 40 Years*, 154.

93 *Alcoholics Anonymous Comes of Age*, 70.

94 *Bill W.: My First 40 Years*, 132–33.

On Evolution

wives, were adherents of the Oxford Group, and were tuned into a similar spiritual wavelength.

During his talk with Bill W. in 1934, Ebby T. had put himself in the difficult position of having to speak to the unconverted as well as to a man who was drinking during the conversation. According to Wing, Ebby was upset that he was not given more credit for his work,[95] and perhaps justifiably so. While Raphael introduces the notion of synchronicity or meaningful coincidence for the meeting between Bill W. and Dr. Bob,[96] this is something that might be applied to the entire chain of events, including the important meetings between Bill W. and Ebby T.

For the task that lay ahead, Ebby was not the determined and ambitious visionary that Bill W. became; his chum had difficulty staying sober. 'PASS IT ON' says that Ebby had not read James's *The Varieties of Religious Experience* himself, though it was recommended by other Oxford Group members.[97] Bill W. studied the book closely, gleaning ideas and the common denominators of stories that would influence A.A. so much. If one believes that each person has a unique, positive purpose[98] - sadly sometimes, or often, unrealized - it may have been Ebby T.'s to "carry the message"[99] and Bill W.'s to cofound Alcoholics Anonymous with Dr. Bob.

After meeting fellow Vermonter, Dr. Bob, Bill W. knew that he had found a partner in the quest to help other alcoholics achieve sobriety. Soon the quest was embodied in Alcoholics Anonymous, the first fellowship or society of the Twelve Step movement. Dr. Bob knew that Bill W. had helped him immensely, allowing the latter to keep the bragging rights

---

95 Nell Wing, *Grateful to Have Been There*, (Park Ridge, IL: Parkside Publishing Corp., 1992), 29.
96 Matthew Raphael, 102.
97 *'PASS IT ON ': The story of Bill Wilson*, 124.
98 Elder James Carpenter said "It has already been planned by the Creator that each person has his or her purpose in this life." (Peter Kulchyski et al, eds., *In the Words of Elders: Aboriginal Cultures in Transition*, [Toronto: University of Toronto Press, 2003], 224).
99 See A.A.'s Twelve Steps at this book's beginning.

belonging to "America's Number One Drunk." Bill W. acknowledged that there were many contributors to the founding and building of A.A.,[100] and gives Ebby T. credit in many places, as noted above, and elsewhere, for example, in his description of the principles of Twelve Step service work.[101]

The Oxford Group became the parent of Alcoholics Anonymous in the late 1930s and played a large role in the founding of A.A.; indeed, it is clear that without the Oxford Group, A.A. might not exist today. Bill W. said that the "Oxford Groupers had clearly shown us what to do. And, just as importantly, we had also learned from them *what not to do* as far as alcoholics were concerned...Our debt to them...was and is immense."[102]

Leading the way, the Oxford Group had key members involved with alcoholism, Rowland H., Ebby T., Bill and Lois W., and Dr. Bob and Anne S. In an almost organic process, Rowland H. saw Carl Jung who gave him guidance; Rowland H. aided Ebby T., and Ebby T. assisted Bill W. who had been kept alive by his wife, Lois W., in New York. In Akron, Bill W. telephoned and met Henrietta Seiberling, who called Anne S., who had kept Dr. Bob alive.[103] The two women arranged a meeting between the two men, and a month later, on Dr. Bob's dry date, June 10, 1935, Alcoholics Anonymous was born.[104]

"Flying Blind": The Oxford Group to A.A.

Alcoholics Anonymous (the Big Book), the fount of A.A. wisdom, explicitly states that A.A. is not religious in any way,[105] outgrowing its Christian

---

100 *The Language of the Heart*, 108, 387.

101 *Alcoholics Anonymous Comes of Age*, 140–41.

102 Ibid, 74-75.

103 Without help from their wives, Bill W. said, he and Dr. Bob would not have lived to create A.A. (*As Bill Sees It: The A.A. Way of Life,* [New York: A.A. World Services, Inc., 1967], 67).

104 A.A. would not use the name Alcoholics Anonymous until 1938.

105 *Alcoholics Anonymous* (The Big Book), xiv.

On Evolution

Oxford Group roots to embrace all faiths.[106] However, this does not stop those who seem convinced otherwise. Katz et al, for example, err in saying that Twelve Step groups have "strong religious biases."[107] Moreover, calling A.A. a "variant" of Christianity as late as 2008, Alexander does not describe A.A.'s evolution to its present-day eclectic spirituality.[108]

However, A.A. and other Twelve Step organizations still use Christian prayers, the Lord's Prayer, for example, and the St. Francis Prayer, Bill W.'s favourite.[109] This can be misleading to those who have spent little time at Twelve Step meetings, or who have made only a cursory reading of Twelve Step literature. I witnessed the reaction of a newcomer to the recitation of the Lord's Prayer at one meeting's conclusion: she immediately left the room and I never saw her again.

In 1935, at the beginning of A.A.'s uncertain "flying-blind period,"[110] the Twelve Step founders had only the Oxford Group and its principles, the Bible, William James, and the ubiquitous coffee pot. Nevertheless, just four years later A.A.'s Twelve Step program had emerged in *Alcoholics Anonymous,* the Big Book, the story of how alcoholics could help each other to achieve sobriety, sanity, and serenity. Now in its fourth edition, the perhaps divinely inspired[111] Big Book remains the central text of Alcoholics Anonymous, and, arguably, of the entire Twelve Step movement. It is sometimes recommended reading in other Twelve Step societies.

Though the Twelve Step pioneers did not use the term mutual aid at that time, it was built into the movement from the start, with alcoholics helping each other in mutuality. Bill W. summarized this early period

---

106 *As Bill Sees It: The A.A. Way of Life,* 34.
107 Alfred Katz et al, 1976, 39.
108 Bruce Alexander, *The Globalization of Addiction: A Study in Poverty of the Spirit,* (New York: Oxford University Press, 2008), 294-300.
109 *'PASS IT ON': The story of Bill Wilson,* 404–06.
110 Alcoholics Anonymous, *Dr. Bob and the Good Oldtimers,* 96.
111 It is quite common for Twelve Steppers to revere the Twelve Step pioneers and their words. Keating writes of divine inspiration in A.A. (Thomas Keating, 2009, 178–207).

in *Alcoholics Anonymous Comes of Age*,[112] and *'PASS IT ON'* describes it in chapters 8 through 10.[113] However, it would take a good fifteen years of hard work to fashion the bare threads of the program into the strong Higher Powered Mutual Aid fabric that many could wear.[114]

The pioneers became convinced that they had found the key to sobriety, sanity, and serenity, by 1953 saying with confidence that much of the guidance contained in A.A.'s *Twelve Steps and Twelve Traditions* might be applied outside of A.A.,[115] predating the growth in the kind and number of Twelve Step societies. Similar words in Al-Anon's first book, *The Al-Anon Family Groups*, written in 1955 with the help of Bill W.,[116] expressed the view that the Twelve Steps could be applied to problems beyond those of the alcoholism for which they were initially designed.[117] Says Al-Anon's first daily reader, the Twelve Step program can be applied to almost all problems, related to alcoholism or not.[118]

As Makela et al's research shows, Twelve Step societies have grown into a worldwide movement for many problems other than alcoholism. Only 5 of the first fourteen Twelve Step fellowships were established to deal with substance abuse: these are Alcoholics Anonymous (1935), Narcotics Anonymous (1953), Overeaters Anonymous (1960), Cocaine Anonymous (1982), and Nicotine Anonymous (1985).[119]

This is important. Most think of Twelve Step programs only as a bulwark against addictions and substance abuse. Just as the Green Party is not only about the environment, Twelve Step programs are not only

---

112 *Alcoholics Anonymous Comes of Age*, 71–75, 138–63.
113 *PASS IT ON': The story of Bill Wilson*, 151–89.
114 The word mutual and its use by Bill W. is explored in more detail in Appendix A.
115 *Twelve Steps and Twelve Traditions* (New York: A.A. World Services, Inc., 1953), 15.
116 Al-Anon Family Groups, *The Al-Anon Family Groups: Classic Edition*, (Virginia Beach, VA: Al-Anon Family Group Headquarters, Inc., 2000), 3.
117 Ibid, 28.
118 Al-Anon Family Groups, *One Day at a Time in Al-Anon*, (Virginia Beach, VA: Al-Anon Family Groups Headquarters, Inc.), 65.
119 Makela et al, 1996, 217.

On Evolution

about substance abuse. Twelve Step programs are also extremely useful as a means of:

- Using effective, non-professional "listen and talk" therapy.[120]
- Discovering one's purpose in life.
- Developing and achieving one's potential.
- Adopting a new moral way of living that includes self-care.[121]
- Learning a philosophy of life that emphasizes spirituality and cooperation, rather than religion and competition.

The Twelve Step founders had unearthed the bones of an ancient, healthier way for human beings to deal with the problems of living, with alcoholism first, and making a life together with others - mutual aid. Bill W. and the pioneers were keen students of human history, learning lessons from movements that had gone before. They then put these lessons into practice in ways that would stand the test of time.

Genius, thorough, and inspired man that he was, and dedicated to knowing how things worked, from making a boomerang in his youth, to science and stock market operations later on, Bill W. studied widely to become the man he wanted to be, and the man that his wife Lois W. expected him to become. According to "Bill's Story," when business was booming in the 1920s, businessmen and financiers were his heroes.[122]

Yet eight pages later, he recorded that influences on him included intellectuals and evolutionists.[123] This is critical, for Bill W. and the pioneers learned that their use of mutual aid, helping one another to gain

---

120 The therapeutic aspect of A.A. was recognized early on. (e.g., David Pittman et al, eds., *Society, Culture, and Drinking Patterns*, 1962), 572–85.

121 Marion Woodman states "[T]here's no moral law anymore," omitting the morality developed in HPMA societies. (Nancy Ryley, *The Forsaken Garden: Four Conversations on the Deep Meaning of Environmental Illness*, [Wheaton, IL: The Theosophical Publishing House, 1998], 72).

122 *Alcoholics Anonymous* (The Big Book), 2.

123 Ibid, 10.

and maintain sobriety, serenity, and sanity, coincided with the theory, from Russian prince, geographer, and anarchist, Peter Kropotkin, that mutual aid is a larger factor of evolution than Darwinian competition. However, the pioneers made a crucial refinement to mutual aid theory - making it Higher Powered - in order that the A.A. and Al-Anon would be the success that they remain to this day.[124]

The Twelve Traditions and the Russian prince

Following the publication of the Big Book and the Twelve Steps in 1939, Bill W. and the pioneers spent much time sifting through their hands-on experience in the growing number of A.A. groups. They also examined the ruins of other movements such as the Washingtonian Society, an over 500,000-member temperance movement begun in the 1850s. His intention? To establish a general set of workable group guidelines for A.A., the Twelve Traditions.[125] This process is recorded in "The Shaping of the Traditions," the first section of Bill W.'s articles in the A.A. Grapevine's *The Language of the Heart* (1988), written between August 1945 and November 1949.

In September 1945 Bill W. wrote "'Rules' Dangerous but Unity Vital," briefly setting out how A.A. might succeed without rules because the program was based on spirituality.[126] It took a good long while to get agreement on the Twelve Traditions, but when A.A. finally ratified them in 1951, ultimate authority was centred only in Higher Powers.[127] This article also contains what may be Bill W.'s first published references to "anarchists," "anarchy," and "mutual aid," each word or term used

---

124 We see later that the overall success of A.A. and Al-Anon, in terms of membership size, is limited.

125 *Alcoholics Anonymous Comes of Age*, 124–25.

126 *The Language of the Heart*, 1988, 6-9.

127 In A.A., newcomers can be in rough shape and sponsors may need to be authoritarian, initially; but ultimate authority is reserved for God. (See A.A.'s Step Two at the beginning of the book.)

twice; here he also says that alcoholics are "true anarchists at heart."[128] Discerningly, Bill W. said further that the Washingtonian Society had used mutual aid to help in its battle against alcoholism, but had underestimated the power of alcohol.[129]

Ten months later, in July 1946, Bill W. again employed "anarchy" in the article, "The Individual in Relation to AA as a Group;"[130] here it is used to describe the lack of rules in A.A.'s Higher Powered Mutual Aid. A third article in January 1947, "Will AA Ever Have a Personal Government?" expands upon A.A.'s unique Higher Powered approach, where authority is spiritually centred, not human centred.[131] Higher Powers hold authority in A.A., Al-Anon, and First Nations societies, outside the human sphere, a fundamental principle of all such societies.

Kurtz reported that, in about 1950, Bill W.'s psychiatrist, Dr. Harry Tiebout, wondered how A.A. could ever hope to succeed given its "self-embraced and self-defining anarchy."[132] The word "anarchy" by itself, then and now, has gotten a lot of bad press, and Bill W. used it at times in *The Language of the Heart* to provide a contrast to A.A.'s benign and beneficial Higher Powered Mutual Aid that had no rules. Repeatedly, Bill W. emphasized that authority in A.A. lies with benevolent and loving Higher Powers, as expressed in the "group conscience" of Tradition Two.[133]

With A.A.'s Twelve Traditions finally approved, Bill W. wrote *Twelve Steps and Twelve Traditions* (1953) to provide more guidance for A.A. members and groups. Here he contrasts complete anarchy[134] with the

---

128 *The Language of the Heart: Bill W.'s Grapevine Writings*, (New York: The A.A. Grapevine, Inc., 1988), 7-8.
129 Ibid
130 Ibid, 32-33.
131 Ibid, 40.
132 Ernest Kurtz, *Not God: A History of Alcoholics Anonymous*, 1979, 128.
133 The "group conscience" (group decision) is best learned about as a decision-making tool by attending Twelve Step meetings and through reading Twelve Step literature. See, for example, A.A.'s pamphlet, *The A.A. Group . . . Where It All Begins*, revised edition, 2005.
134 Alcoholics Anonymous, *Twelve Steps and Twelve Traditions*, 129.

benign and beneficial anarchy of A.A., and elsewhere in the book briefly describes the principles of A.A.'s benign anarchy.[135] In the foreword, for example, he asks why no A.A. member can have authority over another and why government is not applicable in the case of A.A.[136] Bill W. goes on to answer these key questions in the book's discussion of the A.A. Traditions.

How did Bill W. find application for the word anarchy and term mutual aid in the construction of A.A. principles? The pioneers knew, from experience and research, that A.A. must gain some acceptance from mainstream society in order to be successful. Using the Bible, drawing spirituality from religion, and employing the ideas of Dr. Silkworth, James, Jung, and Kropotkin, the A.A. founders were able to design a Twelve Step model of Higher Powered Mutual Aid that would thrive and grow.

In 1946, Bill W. used the word "ancient" to describe A.A. principles that were described earlier in the 1939 Big Book.[137] Many of these principles were from religion and medicine, important wells of support for A.A., especially in the early years, and continuing today. But also in *Alcoholics Anonymous Comes of Age*, Bill W. referred to Alcoholics Anonymous as a "seeming anarchy [with] no human authority whatever,"[138] and further along as a "benign anarchy . . . with a greater personal freedom than any other society knows. We cannot be *compelled* to do anything."[139]

In 1953, Bill W. discussed why A.A. has no government organization.[140] In a talk he gave at A.A.'s 1957 General Service Conference, he also

---

135 Ibid, 16, 132.

136 Ibid, 16. Lois W. cited this in her discussion of the Al-Anon Twelve Traditions in *The Al-Anon Family Groups, Classic Edition* (2000) as "obedience to the unenforceable."166. Her phrase is further noted in Al-Anon's *Many Voices, One Journey* (Virginia Beach, VA: AFG Headquarters, Inc., 2011), 386.

137 *The Language of the Heart*, 21.

138 *Alcoholics Anonymous Comes of Age*, 105.

139 Ibid, 224.

140 Alcoholics Anonymous, *Twelve Steps and Twelve Traditions*, 172–75.

On Evolution

referred to A.A.'s benign anarchy. And in a 1958 letter, flashing the brilliance that perhaps he alone was capable of, Bill W. indirectly contrasted the members of A.A.'s benign anarchy with those who live under dictatorships and hierarchical human authority.[141] Because it is a natural phenomenon, rooted in the innate powers of the Earth itself, HPMA works immeasurably better than artificial authoritarian hierarchy.

The disaster of World War II and the struggle for world supremacy were evident to Bill W.[142] But at the beginning of the Cold War, his prudence came to the fore, and he avoided discussion of the benign anarchy of A.A. in *Twelve Steps and Twelve Traditions* (1953), waiting to describe it more explicitly in *Alcoholics Anonymous Comes of Age* (1957). In the latter book, he refers to the man who is the source of his ideas on mutual aid and anarchy, ideas that are central to the foundation of the Twelve Traditions and to the Twelve Step movement. This man is none other than the "gentle Russian prince,"[143] the long-overlooked geographer and anarchist, Peter Kropotkin, author of *Mutual Aid: A Factor of Evolution* and many other books.

At first, Bill W. says in a disarming manner that there is a greater freedom in Alcoholics Anonymous than in other societies. He calls A.A. anarchist because A.A. members, as noted, "cannot be *compelled* to do anything," but associate with each other voluntarily in a common interest - dealing with the single issue of alcoholism. Bill W. was careful to differentiate between the benign anarchy of A.A., "an association of the benign sort the prince [Kropotkin] envisioned,"[144] and the anarchy of those who believed only in violence as a means of achieving their ends.

In *Alcoholics Anonymous Comes of Age*, Bill W.'s early choice of words was cautious, too, calling A.A. only a "seeming anarchy" and saying that A.A. would be an "irresponsible anarchy" without its services.[145] Following this,

---

141 *'PASS IT ON': The story of Bill Wilson*, 373.
142 *Twelve Steps and Twelve Traditions*, 130.
143 *Alcoholics Anonymous Comes of Age*, 224.
144 Ibid, 225.
145 Ibid, 140.

he said that Twelve Step ideas were borrowed from Kropotkin's anarchy, among other political systems, in the fashioning of A.A.[146] White says that A.A. is a "spiritually driven anarchy."[147]

Again referring to the development of the Traditions in *Alcoholics Anonymous Comes of Age*, Bill W. closely associated them with "self-sacrifice" for the "common welfare," without mentioning Kropotkin, perhaps a source of these words as well. Later still, the single reference to the prince in *Alcoholics Anonymous Comes of Age* became more obscure. In *As Bill Sees It*, Kropotkin is described simply as an "idealist."[148]

The Twelve Step pioneers did not go into detail about the origins of the benign anarchy of A.A. As the chief architect of A.A., Bill W. undoubtedly knew that, during the Cold War of the 1950s, full acknowledgement of a debt to Kropotkin might have been the kiss of death for his child A.A.[149] But Bill W. had given some credit for A.A. principles of benign anarchy to the Russian prince.

More open recognition of Peter Kropotkin might be appropriate, it is suggested, to increase the worldwide appeal of Higher Powered Mutual Aid, and to assist in the "change the self/change the world" work that the Twelve Step pioneers have begun.[150] In this vein, the prudence shown by the founders in not crediting Kropotkin more directly might now be countered with the courage to do so. This would be a positive step forward in broadening, deepening, and strengthening the foundations of the Twelve Step movement, in a world that desperately needs its help.

Bringing Kropotkinist mutual aid theory out of the shadows of Twelve Step history - and I daresay world history - is furthermore necessary to relegate the destructive parts of Darwinism to those very

---

146 Ibid, 225–26.

147 William White, *Slaying the Dragon*, 153.

148 *As Bill Sees It: The A.A. Way of Life*, (New York: A.A. World Services, Inc., 1967), 50.

149 Matthew Raphael says that "A.A. was shaped by cold war ideology no less than any other American institution..." (*Bill W. and Mr. Wilson*, 145).

150 Matt Ridley gives full marks to prince Kropotkin in his book, *The Origins of Virtue: Human Instincts and the Evolution of Cooperation*, (New York: Viking, 1996).

On Evolution

same shadows. The Twelve Step founders knew that their own societies could not withstand the destruction of the wider world that unbounded, competitive, "survival of the fittest" Darwinism has unleashed upon us. "Survival of the fittest" was first coined by Herbert Spencer in 1864, and Darwin made use of it to mean the survival of the strongest, the most competitive. It was left to Kropotkin to describe the fittest as those who cooperate best, who are most sociable, and who practice mutual aid. We explore this in the next chapter.

The reference to the Russian prince in *Alcoholics Anonymous Comes of Age* is a very important one. Upon examining Kropotkin's *Mutual Aid: A Factor of Evolution*, possibly in the early to mid-1940s, Bill W. might have used a phrase he favoured: "Another ten-strike!" Given what we know of Bill W., his unparalleled influence in A.A., and his awareness of that influence, the brief reference to the Russian prince in *Alcoholics Anonymous Comes of Age* would be all that was needed to lead us to Kropotkin.

After all, why would Bill W. and the founders refer to the Russian prince at all, except in the hope that a more opportune moment might present itself to expand upon the place of benign anarchy in the evolution of A.A. and the Twelve Step movement? Noted earlier was Bill W.'s hope that pioneering in A.A. would never end.[151] The time is ripe for more ground-breaking, a deeper examination of Twelve Step history and HPMA's vast, still largely untapped potential for good in the world.[152]

Over 50 years after the publication of *Alcoholics Anonymous Comes of Age*, the accomplished Russian does not seem to have been identified anywhere else. In a letter sent in October 2000, the A.A. Archives was asked if the Russian prince of *Alcoholics Anonymous Comes of Age* was in fact Peter Kropotkin. Unfortunately, the Archives was only able to reply that they had no record of Kropotkin in their files.[153]

In writing his book about A.A., Ernest Kurtz may have been trying to deflect attention away from the subject of anarchism. He renames Bill

---

151 *Alcoholics Anonymous Comes of Age*, 80.
152 This is addressed later on in the book.
153 Letter, November 6, 2000.

W.'s benign anarchy, "joyous pluralism - the reality of needing others in enriching mutuality,"[154] and only mentions Kropotkin and *Mutual Aid* in an endnote, where he suggests a "careful reading" of Peter Kropotkin's book.[155]

Bufe is one student of Alcoholics Anonymous to note the extensive use of anarchist principles in A.A. He observed that A.A. is organized on "non-coercive"[156] democratic lines, "similar to organizational models developed by anarchist theorists."[157] Using Kurtz's technique - citing anarchist theorists in footnotes, including Kropotkin's *Fields, Factories and Workshops*, but not *Mutual Aid* - Bufe fails to describe exactly what Bill W.'s role was here, and which anarchist theorists might have been used to make contributions to A.A. organization and structure. A close reading of *Alcoholics Anonymous Comes of Age* will find the Russian prince waiting.

Who Was Peter Kropotkin?

So who was the Russian prince, a man who remains obscure in the Twelve Step movement, and in the world, perhaps even more obscure than Bill W. is outside the Twelve Step movement? Paul Avrich says that Kropotkin was well known in the United States and Canada in his visits in 1897 and 1901. Over a forty-year period, he writes, Kropotkin was very interested in the United States, and that, of all the Russian visitors who came to North America at the turn of the century, Kropotkin "made the greatest impression."[158] Given Bill W.'s wide reading and interest in mutual aid, evolution, and social architecture, Kropotkin would not have escaped his attention.

---

154 Ernest Kurtz, *Not God: A History of Alcoholics Anonymous*, (Center City, MN: Hazelden Educational Services, 1979), 220.
155 Ibid, 330, endnote 40.
156 Charles Bufe, *Alcoholics Anonymous: Cult or Cure?* 58.
157 Ibid, 79.
158 Paul Avrich, *Anarchist Portraits*, (Princeton: Princeton University Press, 1988), 79.

On Evolution

The prince is described as a gentle man, too, by Woodcock et al, in their biography *From Prince to Rebel*. While gentleness was a prominent facet of his character, he was as complex a man as any, and openly confessed to forceful attitudes and behaviour at certain times. Kropotkin said, in his *Memoirs of a Revolutionist* (1899), "Having been brought up in a serf-owner's family, I entered active life, like all young men of my time, with a great deal of confidence in the necessity of commanding, ordering, scolding, punishing, and the like."[159] However, he realized later that such authoritarian methods work "admirably in a military parade, but [are] worth nothing where real life is concerned…"[160]

In *Memoirs*, Kropotkin describes a boyhood made up of near equal parts of study, roughhousing, and culture. At the age of 20, from his vantage point in the Corps of Pages in the court of Czar Nicholas I, the prince witnessed the beginnings of a great fire, took control, and demonstrated leadership, heroism, and the use of "sheer force"[161] to persuade others to action. Later, in his mid-30s, after a political demonstration in which a clash was avoided, Kropotkin said, "I do not know what feeling prevailed in us most . . . relief at having been spared an undesired fight, or regret that the fight did not take place."[162]

As a young military explorer in Siberia, Kropotkin was interested in furthering Russian imperial aims along the Chinese border. Though Woodcock et al minimize the importance of this,[163] Kropotkin describes it in some detail,[164] obliquely justifying it by referring to "Canadian *voyageurs*"

---

159 Peter Kropotkin, *Memoirs of a Revolutionist*, (New York: Grove Press, Inc., 1968, 1st published 1899), 216.
160 Ibid
161 Ibid, 160.
162 Ibid, 398.
163 George Woodcock et al, *From Prince to Rebel*, 67–70.
164 Peter Kropotkin, *Memoirs of a Revolutionist*, 200-12.

colonization of the Mississippi.¹⁶⁵ He had the notion that north Manchuria was unpopulated and worthy of invasion by Russian settlers.¹⁶⁶

In another instance, with a band of Cossacks accompanying him, Kropotkin was disguised to avoid suspicion during an actual expedition into Manchuria. When confronted by a man he described as an "old half-blind Chinese functionary,"[167] the prince demonstrated deceptive and then authoritarian behaviour in ending the matter, by telling his Cossacks, "Enough of talking...give the order to saddle the horses."[168]

In early 1872, Kropotkin spent three months with "revolutionists" in Western Europe and was inspired by their work before returning home. But with "no possible ground, legal or semi-legal, for . . . a struggle,"[169] his activities nevertheless led to his imprisonment at St. Petersburg in 1874. After his great escape, he landed in England in 1876, six years before the death of Darwin.[170] Kropotkin did not give up his scientific career completely, and supported himself by writing and translating scientific papers.

However, Kropotkin's primary interest remained political activism. As Avrich notes, he was an active revolutionary, "not content to sit at his writing table."[171] He was again imprisoned in 1882, in France, but was freed four years later, partly due to an outcry from international scientists and despite diplomatic pressure from Czarist Russia.[172]

---

165 Ibid, 205.

166 Ibid. This is disturbingly similar to Woodcock et al's use of "empty plains" (George Woodcock et al, 283) to describe Canada's prairies, and Kropotkin's encouragement of Russian emigration to that country. The "empty plains" were the hunting grounds of First Nations peoples before the slaughter of bison that they depended upon, and during the invasion by farmers, fur traders, miners, settlers, soldiers, and police.

167 *Memoirs of a Revolutionist*, 206.

168 Ibid, 205.

169 Ibid, 309.

170 I have found no record of a meeting between Darwin and Kropotkin.

171 Paul Avrich, *Anarchist Portraits*, 73.

172 George Woodcock et al, *From Prince to Rebel*, 191-97.

On Evolution

Yet later in life the prince did not like coercion.[173] His sentiments were clear, for example, regarding the Czarist terrorism practiced in his homeland, described extensively in *Memoirs* and elsewhere. Strongly opposed to the Czarist regime he was brought up under, he became equally opposed to Marxism[174] and to Lenin's brutal state communism, installed after the Russian Revolution of 1917.

In 1901, Kropotkin had simply been dismissive of communism, saying, "We have so often demonstrated that State Communism is impossible, that it is useless to dwell on this subject."[175] In 1905, still predating the Revolution, Kropotkin sometimes became incensed about the matter. In conversation with a Swiss scientist, he "…spoke in harsh terms of Marx, and still more harshly of Engels…His conversation showed warm interest and natural exuberant charm, suddenly interrupted by dire wrath against Marxists…"[176]

Quoting President Woodrow Wilson's representative in Russia, Edgar Sisson, Woodcock et al note that in the immediate post-revolutionary period in Russia, Kropotkin described the Bolsheviks as "…aliens, enemies of Russia, robbers and gangsters, set upon looting and destruction."[177] He thought Lenin "…a madman, an immolator, wishful of burning, and slaughter…willing to betray Russia as an experiment."[178]

But Kropotkin had a sense of humour, too, tending toward self-deprecation. During the Czarist reign, and in his participation in an underground circle, Kropotkin and others pulled a trick on a female comrade, an example of "stupid jokes which men sometimes think funny."[179] His first

---

173 Peter Kropotkin, *Memoirs of a Revolutionist*, 401.
174 Ibid, 386.
175 Peter Kropotkin, "Communism and Anarchy," in *Small Communal Experiments and Why They Fail*, (Petersham, Australia: Jura Books, 1997, first published 1901), 10.
176 George Woodcock et al, *From Prince to Rebel*, 295.
177 Ibid, 406.
178 Ibid, 407.
179 Peter Kropotkin, *Memoirs*, 319.

exposure to the "important matter [of] tea-drinking"[180] in Edinburgh, Scotland, is an amusing and finely written anecdote.

Kropotkin's character seems to have been defined by compassion toward others. After entering the court of the Czar as a youth, he saw that attempts at Russian social and economic reforms were lost amid self-serving pageantry,[181] the "bread and circuses" that were the stock in trade of monarchs and dictators. Later, during geographical explorations in Finland in the early 1870s, Kropotkin turned down a prestigious Russian scientific post, and, after some thought, made an important change of heart to take up the struggle.[182]

It was Kropotkin's compassion and sensitivity toward others that ultimately made him the gentle person that Bill W. estimated him to be, a man committed to both individual freedom and the common welfare. Prince Kropotkin's studies and experience led to the development of optimistic social, economic, political, moral, and environmental ideas in a series of books, including *Mutual Aid*; three inspirational excerpts follow:

> The mutual-aid tendency in man has so remote an origin, and is so deeply interwoven with all the past evolution of the human race, that it has been maintained by mankind up to the present time . . . when even the greatest calamities befell men . . . the same tendency continued to live in the villages and among the poorer classes in the towns . . . it reacted even upon those ruling, fighting, and devastating minorities which dismissed it as sentimental nonsense. And whenever mankind had to work out a new social organization . . . its constructive genius always drew the elements and the inspiration for the new departure from that same ever-living tendency.[183]

---

180 Ibid, 380.
181 Ibid, 141–42.
182 Ibid, 235–41.
183 Peter Kropotkin, *Mutual Aid*, 223.

On Evolution

> [We] are struck with the immense part which the mutual-aid and mutual support principles play even now-a-days in human life. Although the destruction of mutual-aid institutions has been going on in practice and theory, for full three or four hundred years, hundreds of millions...continue to live under such institutions.[184]
>
> [T]he ethical progress of our race, viewed in its broad lines, appears as a gradual extension of the mutual-aid principles from the tribe to always larger and larger agglomerations, so as to finally embrace one day the whole of mankind, without respect to its divers creeds, languages, and races.[185]

Eloquent and hopeful words such as these remain only prophetic. The curtailment of mutual aid in preference to the competitive Darwinist "struggle for existence" has indeed reduced moral progress [186] in the world in the last centuries. Kropotkin said further:

> [T]he need of mutual aid...re-asserts itself again, even in our modern society, and claims its rights to be, as it always has been, the chief leader towards further progress.[187]
>
> In the practice of mutual aid, which we can retrace to the earliest beginnings of evolution, we thus find the positive and undoubted origin of our ethical conceptions; and we can affirm that, in the ethical progress of man, mutual support - not mutual struggle - has had the leading part. In its wide extension...we also see the best guarantee of a still loftier evolution of our race.[188]

---

184 Ibid, 229.

185 Ibid, 224.

186 Ronald Wright laments the lack of "moral progress" in *A Short History of Progress*, (Toronto: House of Anansi Press, 2004), 4, 34.

187 Peter Kropotkin, *Mutual Aid*, 292.

188 Ibid, 300. Echoing Kropotkin, Robinson says, "...mutual aid...is as old as human history." (William Miller et al, eds., *Treating Addictive Behaviors*, [New York: Plenum Press, 1986], 291).

However, Avrich criticizes Kropotkin's belief, faith, and hope in mutual aid, saying:

> Kropotkin's optimistic view of human nature, his faith in mutual aid as opposed to Darwinian competition, his belief that the centralized state had reached its apogee in his own time...have scarcely been borne out in our [20th] century of world wars and large scale government...He took insufficient account of the naked violence that dominates the life of most animals...And because of this he exaggerated the extent of human solidarity in the world and emphasized the bonds rather than the hatreds and divisions."[189]

Avrich is mistaken on three counts. Kropotkin's optimism was partly deliberate. In reflecting on revolutionary writing, he felt that "[I]t is hope, not despair, which makes successful revolutions."[190] Secondly, Kropotkin always recognized that there was competition in evolution; he only believed that mutual aid was a larger factor. Third, "naked violence," if it occurs at all, is far from a dominant activity in the life of any creature, human beings included, despite our apparent fondness for bloodshed.

In the consideration of Kropotkin's ideas by the Twelve Step pioneers, further facets of his character were probably decisive. These are in the areas surrounding authority, spirituality, and religion. The most important difference between Kropotkinist mutual aid and Twelve Step HPMA is in the positioning of authority. Alcoholics Anonymous had its start in the Oxford Group, and through James and Jung, the Twelve Step founders never forgot that the guidance of Higher Powers could lead them to a way out. Bill W.'s spiritual awakening was key, a turning point for him, and - to say the least - for many others.

Twelve Step societies, and as we shall see among First Nations societies, work with the critical idea that authority is centred in Higher Powers, while authority in Kropotkin's brand of mutual aid remained centred in

---

189 Paul Avrich, *Anarchist Portraits*, 75.
190 Peter Kropotkin, *Memoirs*, 418.

On Evolution

human beings, through the various powers of economics, politics, and science. The idea of human limitations did not appeal to Kropotkin; he might have thought that man was God.

Kropotkin had faith in the power of anarchic organization to turn human and natural resources into monetary wealth, but expressed little faith in spiritual guidance. In *Memoirs*, Kropotkin wrote of his early exposure to religion, "Of all that I had ever heard in the church only two things had impressed me: the twelve passages from the Gospels, relative to the sufferings of the Christ [Jesus]…and the short prayer condemning the spirit of domination."[191] Later, he found services in churches "theatrical" and even shocking, "…the moreso when I saw there with what simple faith some retired Polish soldier or peasant woman would pray in a remote corner."[192]

The Russian prince does not appear to have moved beyond his early exposure to religion to faith in his own understanding of a God that he could trust with authority. His understanding and practice of mutual aid was Godless. By contrast, the centreing of authority in a loving God, as expressed in the Twelve Step group conscience, became unquestionably essential to the success of the Twelve Step movement.[193]

So strong was Kropotkin's mutual aid theory that, before his first imprisonment in Czarist Russia, decision-making in underground circles used a group conscience.[194] Kropotkin and others saw the value of this for individual expression within the common welfare. However, Kropotkin failed to replace human authority with Higher Authority, and, admirable as it seemed, his model of mutual aid foundered on the shoals of human ego. He admitted, "In the hundreds of histories of communities which I have had the opportunity to read, I always saw that the introduction of any

---

191 Ibid, 96.

192 Ibid

193 Keating writes of ridding ourselves of a punishing or tyrannical God. (Thomas Keating, *Divine Therapy and Addiction: Centering Prayer and the Twelve Steps*, [Brooklyn, NY: Lantern Books, 2009], 81–82).

194 George Woodcock et al, *From Prince to Rebel*, 128.

sort of elected authority has always been, without one single exception, the point which the community stranded upon..."[195]

The placement of authority remained an obstacle to the success of Kropotkin's work. The Twelve Step pioneers used trial and error a great deal, and extensive research, but came up with improved results. The Twelve Step founders learned lessons here, too. The democratic election of leaders specifically avoids giving them authority: authority resides with Higher Powers as expressed in the group conscience. Twelve Step leaders "...are but trusted servants, they do not govern."[196]

Suspicious of religion, Kropotkin cited it as another reason for the failure of his anarchic communities.[197] In associating anything spiritual or religious with hierarchical authority, he appears to have lost perspective on the subject, unaware of the value of guiding authority that benevolent Higher Powers could provide in mutual aid. The first important difference between Kropotkin's mutual aid, and that of First Nations and Twelve Step societies, was thus the placement of authority in Higher Powers, beyond the reach of those who might speak or act in authoritarian ways.

As demonstrated by Twelve Step and, as we shall see, First Nations societies, mutual aid is made workable when power is placed outside the human sphere, in Higher Authority. Without the imposition of human authority, individuals can practice natural, unfettered, God-given freedom and creativity within the common welfare, without the fear and anger that would inevitably come from the application of human domination and hierarchical authoritarianism.

Related to this last facet of Kropotkin's character were circumstances over which he was truly powerless. After his return home from Europe to engage in the struggle for justice against the Czar, even a man of his

---

195 Peter Kropotkin, "Proposed Communist Settlement: A New Colony for Tyneside or Wearside," in *Small Communal Experiments and Why They Fail*, (Petersham, Australia: Jura Books, 1895, republished 1997), 16.

196 See A.A.'s Tradition Two at the front of the book.

197 Peter Kropotkin, "Communism and Anarchy," in *Small Communal Experiments and Why They Fail*, (Petersham, Australia: Jura Books, 1901, republished 1997), 14.

royal stature, as previously noted, saw "no possible ground, legal or semi-legal, for such a struggle."[198] The right to engage in political meetings was curtailed and then stopped by the Czar.

Kropotkin's initial battle was against the tyranny of the Czars, and he proved powerless over it. With his book, *Mutual Aid*, he took up a perhaps more manageable struggle, countering the prevailing ideology of Darwinian competition with Kropotkinist mutual aid. While both Czarist and Darwinist economic, political, and social regimes can be seen as ill, Kropotkin's fight against them proved still less manageable than A.A.'s battle with alcoholism. The Twelve Step pioneers knew that the entertainment of outside issues, such as democracy, environmentalism, justice, politics, poverty, or religion, would divert them from dealing with alcoholism effectively, perhaps destroying their efforts altogether.

The Russian prince sometimes saw revolution as a means of changing society, as have others,[199] but the Twelve Step pioneers concluded that even democratic society was sick and in need of healing. This second view seems to be the correct one, despite the tendency to find someone else to blame for our ills. The Twelve Step movement has shown that instead of blaming others, it is far wiser to focus on ourselves. By first transforming our individual selves, we might then transform families, communities, nations, and the world, with guidance from the Higher Powers.[200]

The Twelve Step founders confronted another relevant aspect of the disease of alcoholism related to the positioning of authority. This centres on humility. Bill W. learned the hard way to ego-deflation,[201] and struggled with it for much of his life, but, while Kropotkin rejected the

---

198 Peter Kropotkin, *Memoirs*, 309.

199 Still seen as a way forward in the 21st century is violence in Africa, the Americas, Asia, Australia, Europe, the Middle East, and in terrorism around the world.

200 On our topsy-turvy planet, it is sometimes necessary to focus on others first: it was necessary to stop the Axis powers during World War II without addressing underlying illnesses such as alcoholism. Many made the ultimate sacrifice to bequeath us a safer, more peaceful world.

201 Appendix D addresses the subject of ego-deflation, or "right-sizing" the ego.

upper class trappings into which he had been born, he could not shake attitudes of human and scientific superiority. Avrich quotes sources that saw Kropotkin as convinced of the rightness of his own views.[202] As an accomplished scientist of the time, belief in Higher Powers and a humble human place in the world were not options that he seriously considered.

Raised in an authoritarian household and experiencing the rebellion against authority in his lifetime, then imprisoned twice, Kropotkin may even have concluded that individual liberty should be unbounded. He apparently said, in regard to anarchism, "...every effort [should] be made to extend the liberty of the individual in all directions."[203] It is such statements that give anarchy an unenviable reputation; ironically, it reminds me of the "anything goes" morality of the West.[204]

From the vantage point of sheer desperation in the relatively free society of the United States, A.A.'s founders were able to concentrate on themselves to become well. Medicine was sometimes unable to help,[205] and the pioneers came to see that they had to help themselves to build Higher Powered Mutual Aid theory and structure for the single purpose of dealing with alcoholism.[206] In so doing, the Twelve Step pioneers used some of Kropotkin's ideas to strengthen the foundations of Alcoholics Anonymous, and to help alcoholics focus on and overcome the disease of alcoholism. The practical focusing on one's own problems in Twelve Step recovery work is an example to the world, optimistically the beginning of a successful return to the Garden.

---

202 Paul Avrich, *Anarchist Portraits*, 75.

203 Peter Kropotkin, "Communism and Anarchy," in *Small Communal Experiments and Why They Fail*, 14.

204 Abrams criticizes "moral relativism," equating it with our "anything goes" culture. (Nancy Ellen Abrams, *A God That Could Be Real: Spirituality, Science, and the Future of our Planet*, [Boston: Beacon Press, 2015], 129).

205 See, for example, *'PASS IT ON': The story of Bill Wilson*, 100–102, 268, 270, and 305.

206 Bornstein says that A.A. is an example of citizens who have taken matters "into their own hands." (David Bornstein, *How to Change the World: Social Entrepreneurship and the Power of New Ideas*, [New York: Oxford University Press, Inc., 2004], 274).

On Evolution

Yet this much unfortunately endures to the present time: the immense and dark overshadowing of mutual aid by disastrous "me first" attitudes connected to the competitive Darwinist "struggle for existence" and Spencer's "survival of the fittest," represents a serious imbalance in the way we carry out our lives. Its effects lie deeper than we are willing to admit, so blinded are we by the exorbitant investment we have made in current ways of living. Bill W. explored such ideas in 1950s correspondence with a death row prisoner, saying that alcoholics and prison inmates were "... merely the grotesque end products of [society's] own defects."[207]

The views of Bill W. and Peter Kropotkin were similar here, as shown in excerpts from their writings, as follows:

> The whole modern world is in fact coming apart as never before because of political and religious strife; because men blindly pursue wealth, fame, and personal power regardless of the consequences to anyone, even to themselves.[208]
>
> ...the theory...that men...must...seek their own happiness in a disregard of other people's wants is now triumphant all round - in law, in science, in religion.[209]

Kropotkin and Bill W. knew that the state took power away from people, "individual in relation" power and self-regulation. Anarchy here is derived from each one's innate decision making "power from within," not power from outside sources. Innately derived self-responsibility lets us make moral decisions without treading on the rights of others. Individual growth (e.g., in self-regulation, self-responsibility, self-respect...) is stunted in societies with a hierarchy of human authority, thus also stunting the progress of the common welfare. The danger too is that this represents an actual *reversal* of the interests of the individual welfare and the common good, together.

---

207 *'PASS IT ON': The story of Bill Wilson*, 365.
208 *Alcoholics Anonymous Comes of Age*, 233.
209 *Mutual Aid*, 228.

Bill W. and the other pioneers centred authority in Higher Powers to recognize their human powerlessness in the face of alcoholism, one of many illnesses that now affect all of humanity. Peter Kropotkin was aware of the symptoms of illness in World War I, saying that Darwin's "struggle for existence" was used as a "favorite explanation" for war.[210]

---

210 Ibid, preface to the 1914 edition, xxxi.

CHAPTER 3

Darwin and Kropotkin

Two heads are better than one.

SOURCE UNKNOWN

THE TRUE SIGNIFICANCE of Peter Kropotkin's ideas in relation to A.A. and the Twelve Step movement can only be appreciated by referring to the work of Charles Darwin. Kropotkin studied Darwin's books, *The Origin of Species* (1859) and *The Descent of Man* (1871, revised 1874), and concluded that mutual aid was far more important than Darwinian competition. According to Kropotkin, Darwin's competitive "struggle for existence" and borrowed "survival of the fittest" would be better expressed thusly: the fittest are those who are cooperative, who are sociable, and who help each with mutual aid.

Darwin's *The Origin of Species* describes the role of a competitive "struggle for existence" in natural selection, and it remains the most influential tome on evolution. After reading the book, Kropotkin said that, in Siberian scientific explorations, he "[v]ainly looked for the keen competition . . . which Darwin's work had prepared us to expect [but] competition and struggle between higher animals of the same species came very seldom under my notice, though I eagerly searched for them."[211]

Kropotkin recalled that, while he was in England, interpretation of the "struggle for existence" as "Woe to the Weak" was so "deeply rooted

---

211 Peter Kropotkin, *Mutual Aid*, 9.

in this country that it had become almost a matter of religion."[212] This stark reality made it necessary for Kropotkin to take a more diplomatic line toward Darwin's work, so that his own views on evolutionary mutual aid might get a fair hearing.

Though Newman feels that Kropotkin was "vehemently opposed to Darwin's theories,"[213] it seems instead that Kropotkin took what he liked - Darwin's brief mentions of cooperation and sociability - to support his own ideas on speciation, while leaving aside the less palatable parts of Darwin's work.

Muddying the waters, Morris says that Kropotkin was "an avid supporter of Darwin's theory of evolution."[214] Likewise, Alexander thought that Kropotkin was a "resolute" Darwinist.[215] More probably, the prince was *reticent*, loath to share more of his reservations about Darwin's emphasis on competition in natural selection. In the end, Kropotkin wrote, posthumously, 'it must be said that Darwin did not analyse… sociality in animals…to the extent which it deserved…"[216]

Kropotkin and other Russian Darwinists favoured a mutual aid interpretation of Darwin's theory, not the competitive one put forward by European Darwinists.[217] Todes quotes one of Kropotkin's letters as saying, "*We* [Russian Darwinists] see a great deal of mutual aid, *where Darwin and Wallace* [Darwin's compatriot rival] see only struggle."[218]

---

212 Peter Kropotkin, *Memoirs of a Revolutionist*, (New York: Grove Press, Inc., 1968. 1st published. 1899), 499.

213 Ibid, xviii.

214 Brian Morris, *Kropotkin: The Politics of Community*, (Amherst, NY: Humanity Books, 2004), 129.

215 Bruce Alexander, *The Globalization of Addiction*, 2008, 89.

216 Peter Kropotkin, *Ethics: Origin and Development*, (Montreal: Black Rose Books, 1992), 35.

217 Ibid, 137.

218 Daniel Todes, *Darwin without Malthus: The Struggle for Existence in Russian Evolutionary Thought*, (New York: Oxford University Press, 1989), 142.

On Evolution

Russian naturalists, said Kropotkin, differed from those of Western Europe only in certain respects.[219] He cited ideas about mutual aid from the Russian zoologist, Professor Karl Kessler and others,[220] but exaggerated their compatibility with Darwin's work, saying that the primacy of mutual aid over competition was "in reality, nothing but a further development of the ideas expressed by Darwin himself in *The Descent of Man*."[221] These were hopeful words from Kropotkin. But it appears that he was attempting to bridge the gap between the world as he saw it, and the ruthless competitive world that Darwin actually described, in quite unrelenting fashion.

With Darwin's passing in 1881, some of his followers may have taken up the torch of competitive struggle with more zeal than did Darwin. Kropotkin records in his *Memoirs of a Revolutionist* that an 1888 article by Thomas Huxley was so "atrocious"[222] that it became the impetus for the publishing of *Mutual Aid*.[223] Prince Kropotkin pounced upon the opportunity to write about the primacy of mutual aid in evolution.

The term "mutual aid" was adopted by Kropotkin from Kessler, whom he credits for a lecture on the subject in January 1880.[224] In 1879, De Bary had begun using the term "symbiosis"; it included "mutualism," a poor synonym for mutual aid. De Bary's terms were used from then on in the West.[225] Boucher feels that Kropotkin's use of mutual aid, not mutualism, his anarchism, and

---

219 Peter Kropotkin, *Mutual Aid*, 9.

220 Peter Kropotkin, *Mutual Aid*, 6-8. Kessler, said Kropotkin, believed that the "law of Mutual Aid" was "far more important" than the "law of mutual contest." (*Mutual Aid*, xxxviii).

221 Ibid, xxxviii.

222 Kropotkin, *Memoirs*, 499.

223 Kropotkin, *Mutual Aid*, xlii.

224 Ibid, 6-8. Katz et al mistakenly credit Kropotkin for the first use of the term mutual aid. (Alfred Katz et al, *The Strength in Us: Self-help Groups in the Modern World*, [New York: Franklin Watts, 1976], 8).

225 See, for example, Lynn Margulis et al, eds., *Symbiosis as a Source of Evolutionary Innovation*, (Cambridge, MA: The MIT Press, 1991).

his scientific and geographical distance from Europe, hindered acceptance of his work.[226]

Cooperation, Mutual Aid, and Sociability

In the introduction to the first edition of *The Origin of Species*, Darwin mentions his book's limitations, specifically concerning mutual relations; he wrote:

> No one ought to feel surprise at much remaining as yet unexplained in regard to the origin of species…if he makes due allowance for our profound ignorance in regard to the mutual relations of all the beings which live around us…relations [which] are of the highest importance.[227]

He repeats this idea later, referring to "…our ignorance on the mutual relations of all organic beings…"[228] Further, he says that, "Let it be borne in mind how infinitely complex and close-fitting are the mutual relations of all organic beings to each other and to their physical conditions of life"[229] without saying that cooperation, mutual aid, and sociability are *intrinsic* to all life, and not produced by natural selection.

In a few places, Darwin describes mutual relations, for example, among insects (e.g., bees and ants),[230] but does not portray it as mutual aid. Furthermore, these references never detract from his paramount conviction that evolution through natural selection occurs through competition in his "struggle for existence." Then, very near the end of

---

226 Douglas Boucher, ed., *The Biology of Mutualism: Ecology & Evolution*, (London: Crown Helm, 1985), 17-18.
227 Charles Darwin, *The Origin of Species*, 1968, first edition 1859, 68-69.
228 Ibid, 129.
229 Ibid, 130.
230 Ibid, 141-42, 243–63.

On Evolution

The Origin, poetic, almost wistful words appear, describing the mutual interdependence of various forms of life; here Darwin said:

> It is interesting to contemplate an entangled bank, clothed with many plants of many kinds, with birds singing on the bushes, with various insects flitting about, and with worms crawling through the damp earth, and to reflect that these elaborately constructed forms, so different from each other, and so dependent on each other in so complex a manner, have all been produced by laws acting around us.[231]

Such words were, however, only an afterthought by Darwin, "...a sideshow, a detail that had to be explained away" according to Nowak.[232] In view of the limitations of the first edition of *The Origin of Species* about the role of mutual relations in evolution - limitations Darwin admitted - his *The Descent of Man* may have been intended partly as a correction to his earlier emphasis on competition in *The Origin*. Ambiguous references to cooperation, mutual aid, and sociability in *The Descent* seemed to show that Darwin had become aware of the value of mutual aid, and Kropotkin used it to his advantage.

Early in *The Descent*, Darwin seemed to say that cooperation and mutual aid represented a factor of evolution, as Kropotkin maintained. The following are two excerpts taken from that book; in a discussion of human ancestry, Darwin wrote:

> [I]t might have been an immense advantage...to have sprung from some comparatively weak [but social] creature [the chimpanzee, not the gorilla].[233]

---

231 Charles Darwin, *The Origin of Species*, first edition, 459.

232 Martin Nowak et al, *SuperCooperators: Altruism, Evolution, and Why We Need Each Other to Succeed*, (New York: Free Press, 2011), xvi.

233 Charles Darwin, *The Descent of Man*, revised ed., (Chicago: Rand, McNally and Company, 1874), 61. Partly through the work of Fossey, the gorilla has been shown to be a social animal after all, as all creatures are to one degree or another. (Dian Fossey, *Gorillas in the Mist*, [Boston: Houghton Mifflin, 1983]).

> The small strength and speed of man, his want of natural weapons, etc., are more than counterbalanced...by his intellectual powers...and...by his social qualities which lead him to give and receive aid from his fellow-men.[234]

Here Darwin suggests that social qualities lead us to adopt mutual aid, and that sociability springs up naturally as a human impulse "to give and receive aid."

But later in *The Descent*, Darwin alters his language, offering instead that social qualities might actually have been *produced* by natural selection; he states:

> In order that primeval men...should become social, they must have acquired the same instinctive feelings, which impel other animals to live in a body...social qualities... were no doubt acquired through natural selection...[235]

Nonsense. First, Darwin clearly implies that primeval men, at some point, were not social, forgetting what he'd said about social chimpanzees. He says further that for humans to *become* social, they had to acquire "instinctive feelings." These feelings - "social qualities" - he conjectures (above), were "no doubt acquired through natural selection." His writing at this point in *The Descent* leaves one with the impression that he is unable to make up his mind about the origins of cooperation, mutual aid, and sociability.

Furthermore, Darwin does not explain how one might acquire instinctive feelings in natural selection, in this case social qualities, through a struggle that was always competitive. He may have been trying to shore up his theory of natural selection through the competitive "struggle for existence" by allowing a minor and ambiguous role for cooperation, mutual aid, and sociability.

---

234 Charles Darwin, *The Descent of Man*, 61.
235 Ibid, 126.

On Evolution

Darwin's Metaphor, and Contradiction

Early in the introduction to his book, *Mutual Aid*, Kropotkin went to some lengths to credit Darwin for his brief description of the "Struggle for Existence" as a metaphor.[236] Darwin's words in the first edition of *The Origin of Species* were:

> I use the term Struggle for Existence in a large and metaphorical sense, including dependence of one being on another…[237]

Here, Darwin was clear that the struggle for existence included "dependence of one being on another." In *The Origin*, unfortunately, this became a peripheral concern, not occupying a part consistent with full use of his metaphor.

Instead, *The Origin* was devoted overwhelmingly to individual competitive struggle and not to cooperation, mutual aid, and sociability.[238] *The Origin of Species*, says Alexander, depicts evolution "largely as the product of ruthless individual competition…"[239] Darwin wrote little of natural selection through "dependence of one being on another," or through cooperation, mutual aid, and sociability. Competition ruled his explanations of natural selection and the "struggle for existence" almost exclusively.

It is important to point here again to Darwin's faulty reasoning. An unremitting, competitive struggle should logically yield more competitive struggle, not cooperation, and, taken to its logical conclusion, natural selection via competition might actually *eliminate* cooperation, mutual aid, and sociability. His conclusion that these arise out of a competitive natural selection is thus a simple contradiction.

There were six editions of the *Origin* during Darwin's lifetime, spanning almost fifteen years, and it is perplexing that this error was

---

236 Peter Kropotkin, *Mutual Aid*, 1–4.
237 Charles Darwin, *The Origin of Species*, first edition, 116.
238 Bruce Alexander, *The Globalization of Addiction*, 85.
239 Ibid

not noticed. Darwin's introductory remarks, cited above, regarding *The Origin*'s "... profound ignorance in regard to mutual relations..." might have been greatly expanded upon to alleviate the problem.

Darwin was correct in *The Origin of Species*. In omitting thorough consideration of cooperation, mutual aid, and sociability, his ideas provided an incomplete theory of natural selection. However, what Darwin failed to say was that detailed consideration of these elements *would have changed his theory enormously*.

By the time Darwin wrote *The Descent of Man*, the metaphor of *The Origin* was forgotten. Rather, as noted, he concluded that the struggle for existence did not include sociability, only selected it. The following are further excerpts from *The Descent*:

> With strictly social animals, natural selection sometimes acts on the individual, through the preservation of variations which are beneficial to the community.[240]
>
> Animals endowed with the social instincts take pleasure in one another's company, warn one another of danger, defend and aid one another...As [they] are highly beneficial to the species, they have in all probability been acquired through natural selection...[241]
>
> [T]he social instincts . . . may be safely attributed to natural selection.[242]

In the first excerpt above, Darwin says that some animals are "strictly social." This alone contradicts his theory of natural selection through competition: animals that are "strictly social" would be unable to survive the competitive struggle. Darwin then maintains, in the other two excerpts, that because the social instincts are beneficial to the species, they must "in all probability been acquired through" and "safely attributed

---

240 Charles Darwin, *The Descent of Man*, 60.
241 Ibid, 605.
242 Ibid, 612.

to" natural selection. However, if natural selection operates only through competitive struggle - this needs to be repeated - that does not actually include cooperation, how might cooperation, mutual aid, and sociability be selected?

In truth, the "social instinct," yielding cooperation, mutual aid, and sociability, is not a product of natural selection at all, but an *innate* quality of all living things. The unadorned evidence shows that all forms of life are cooperative, inclined to be helpful to others, and sociable. These innate instincts are used for individual, familial, community, country, and global benefit. Conception requires a cooperative getting together of egg and sperm, with some competition thrown in for good measure. Procreation of even the most basic life forms requires cooperation, mutual aid, and sociability that is not selected for. Specifically, rather, they are intrinsic qualities of all living species.

At the very tiny level, molecules are physically supported by, and are supportive of, other molecules - mutual aid (or mutual support) - in air, land, water, vegetation, other animals, and all the animate and inanimate of Creation. If molecules did not support each other in this way, we, the world, and perhaps the Universe, would collapse in an instant. Seen in this light, competition is an infinitesimally small enterprise.

That mutual aid might have been a factor of evolution did not impress Darwin and his view of the world. Instead, what is clear was his single-minded determination to describe natural selection as occurring only through the Malthusian "law" of unforgiving competitive struggle. Very influential with Darwin, and among European Darwinists, was T. R. Malthus's "Essay on the Principle of Population," an "...account of humanity outstripping its food supply, and the weak and improvident succumbing in the struggle for the available resources."[243]

Halfway through the first edition of *The Origin*, Darwin put forward:

---

243 Adrien Desmond et al, *Darwin*, (London, UK: Michael Joseph, 1991), 264.

... one general law, leading to the advancement of all organic beings, namely, multiply, vary, let the strongest live and the weakest die.[244]

Morris says that Kropotkin admired Darwin, but felt that "Malthus's influence had...led Darwin to certain erroneous conclusions."[245] In this context, it is seldom observed that Malthus was not a naturalist at all, but a clergyman. Darwin had seriously entertained becoming a clergyman, too, before his voyage aboard the *Beagle*.

In the conclusion to *The Origin of Species*, Darwin made no mention of the role of cooperation, mutual aid, and sociability in a natural selection that was reduced to the "war of nature, from famine and death."[246] Boucher says that, though Darwin spent much time studying mutualism - Kessler's and Kropotkin's mutual aid - his fundamental explanation was natural selection by competition.[247] Biology has a "dark side," says Nowak, the "shadowy aspect of nature" that Darwin favoured, saying, "[The] fittest win this endless 'struggle for life most severe' and all others perish."[248]

Todes offers an excuse for Darwin's excesses, saying that he "had been preoccupied with the consequences of individualistic competition and so had never pursued his insight about mutual aid."[249] Darwin took over 20 years to write *The Origin of Species*, and over 30 years for *The Descent of Man*, so perhaps he had the time but not the inclination.[250]

---

244 Charles Darwin, *The Origin of Species*, 1968, first published 1859, 263. Rather, "multiply, vary, help each other to live together, then die."
245 Brian Morris, *Kropotkin: The Politics of Community*, 2004, 138.
246 Ibid, 459.
247 Douglas Boucher, *The Biology of Mutualism*, 1985, IV.
248 Martin Nowak et al, *SuperCooperators*, 2011, xi.
249 Daniel Todes, *Darwin without Malthus: The Struggle for Existence in Russian Evolutionary Thought*, (New York: Oxford University Press, 1989), 133.
250 Among other explanations, British hegemony was perhaps too important to be derailed by a cooperative, mutual aid theory of evolution.

On Evolution

In *Darwin*, Desmond et al state that David Livingstone's observations in Africa did not support Darwin's competitive "struggle for existence."[251] Yet even Livingstone's views were trampled upon in the rush to accept an evolutionary theory that supported the social, economic, exterminatory, and imperialist designs of the Europeans of the time. Browne notes that Darwin's theory "dovetailed with...the commercial, entrepreneurial spirit of unfettered competition..."[252]

This is what Kropotkin feared, an interpretation of Darwin's theory that justified "both the European domination of the rest of the world and the predominance of laissez-faire capitalism."[253] *The Cambridge History of the Native Peoples of the Americas* states that Darwin's:

> [T]heory of evolution through natural selection was redeployed by social theorists to naturalize racial and class inequalities and rationalize the exercise of violence and domination in colonialist and capitalist ventures.[254]

This "Social Darwinism," held Himmelfarb, "has become a portmanteau of nationalism, imperialism, militarism, and dictatorship...," and is commonly understood "as a philosophy of extreme individualism, of laissez-faire economics and government."[255] So broad is its sweep that, "From Albania to Alabama, from Russia to Rwanda, Darwin's theories have been used to justify racial conflict and ethnic cleansing."[256]

---

251 Adrien Desmond et al, 1991, 480.

252 Janet Browne, *Charles Darwin: Voyaging. A Biography*, (New York: A. Knopf, 1995), 390.

253 Brian Morris, *Kropotkin: The Politics of Community*, 2004, 134.

254 Frank Saloman et al, eds., *The Cambridge History of the Native Peoples of the Americas*, Vol. III, South America, Part 2, (Cambridge: Cambridge University Press, 1999), 575, note 3.

255 Gertrude Himmelfarb, *Darwin and the Darwinian Revolution*, (New York: W.W. Norton & Company Inc., 1962; first published 1959), 416-18

256 Adrien Desmond et al, *Darwin's Sacred Cause: Race, Slavery and the Quest for Human Origins*, (London, UK: Allen Lane, 2009), xix.

Ashley Montagu fired a broadside at competition in natural selection in *Darwin: Competition and Cooperation*. However, he seems not to have made any more of a dent in it than Kropotkin, to whom he dedicated his book.[257] Montagu wrote:

> What is now quite clear is that Darwin's conception of [the evolutionary] process perfectly fitted the pattern of Victorian social thought and practice. What Darwin's view of life did was to give that social thought and practice a biological validation, a scientific foundation which it had hitherto lacked.[258]

Further removed than Kropotkin from early discussions and writing about natural selection, Montagu was less tentative in his criticism of Darwin; he wrote, too, that:

> In a machine age, Darwin gave a mechanical explanation of evolution. In an age characterized by industrial competition in which no quarter was given, Darwin gave an explanation virtually entirely in terms of competition, in terms of the struggle for . . . existence, the survival of the fittest.[259]

Montagu said also:

> Had Darwin not been so much under the influence of the "struggle for existence"...he might have glimpsed the truth that no animal is...solitary...social life is characteristic of all living things [and] that [we] depend not upon competition, but...cooperation.[260]

---

[257] Ashley Montagu said that Kropotkin's *Mutual Aid: A Factor of Evolution* is "one of the world's great books." (Brian Morris, *Kropotkin: The Politics of Community*, 2004, 135).

[258] Ashley Montagu, *Darwin: Competition and Cooperation*, (New York: Henry Schuman, 1952), 18.

[259] Ibid, 31.

[260] Ibid, 74.

On Evolution

As early as 1938, geographer Lewis Mumford said that Kropotkin's *Mutual Aid* was "the beginning of the more positive study of human ecology,"[261] what might be termed "Kropotkinism." Opined Rene Dubos in 1959, Peter Kropotkin was strongly of:

> "[t]he view that co-operation and mutual help, rather than competition, had been the principal agents of evolution."[262]

In 2011, Nowak et al used science to explore the real meaning and efficacy of cooperation in evolution.[263] So did Abrams and Nowinski in 2015.[264] In doing so, Nowak et al had praise for Darwin's work re: natural selection through competition, but say that this is only part of the story. Following the lead of Kropotkin in *Mutual Aid*, cooperation, say Nowak et al, is even more important than competition in the "struggle for existence," and is the "…principal architect of 4 billion years of evolution."[265]

Commenting that the cornerstones of Darwin's work - mutation and natural selection through competition - are insufficient, Nowak et al say, "…we need a third ingredient, cooperation, to create complex entities, from cells to societies."[266] Briefly mentioning Kropotkin's *Mutual Aid*, Nowak et al emphasize the prince's explanation that cooperation, mutual aid, and sociability are included as a *part of* natural selection, and to Kropotkin, "…is far more important than the law of mutual contest."[267]

---

261 Lewis Mumford, *The Culture of Cities*, (New York: Harcourt, Brace and Company, 1938), 302.
262 Rene Dubos, *Mirage of Health: Utopias, Progress, and Biological Change*, (New Brunswick, NJ: Rutgers University Press, 1959), 63.
263 Martin Nowak et al, *SuperCooperators*, 2011.
264 Nancy Ellen Abrams, *A God That Could Be Real*, (Boston: Beacon Press, 2015); Joseph Nowinski, *If You Work It, It Works!* (Center City, MN: Hazelden Publishing, 2015).
265 Martin Nowak et al, *SuperCooperators*, 280.
266 Ibid, 14.
267 Peter Kropotkin, *Mutual Aid*, xxxviii.

Nowak et al write, however, that only *some* people are predisposed to cooperate,[268] when the truth is that all creatures are so predisposed, obviously some more than others. Michael Tomasello, author of *Why We Cooperate*, explains how social behaviour is innate in us, using young children as an example.[269] We did not evolve into cooperative behaviour, we are born social cooperators, so learning about further cooperation, sociability, and mutual aid has a solid base upon which to build.

Abrams is more circumspect, choosing not to touch upon Darwinian competition, using "violence" instead of cooperation's opposite, competition.[270] Alexander states that "the social instinct and the moral sense [arose] through the course of evolution," but proceeds to sit on the fence, on the one hand saying that the "social instinct" is innate, which it is, and on the other that it is an adaptation,[271] which it is not.

In *A Cooperative Species: Human Reciprocity and its Evolution*, Bowles writes, "No 'gene for cooperation' has been discovered."[272] Nevertheless, he clearly comes down on the side of cooperation as the leading factor of evolution. Ridley said in 1996, that cooperation is instinctive; it is "literally in our nature."[273] He cites examples such as Hutterite communities, ants, and bees, all of which illustrate that the theory of evolution plays out best in cooperative, mutual aid, and sociable behaviour.

In its cover story, the August 2015 edition of *Scientific American* describes a hypothesis by Curtis Marean. He writes about the primacy of cooperation in successful human evolution, for example, in our ability to

---

268 Martin Nowak et al, *SuperCooperators*, 90.

269 Michael Tomasello, *Why We Cooperate*, (Cambridge, MA: The MIT Press, 2009). Tomasello says of children: "[for whom sociability is natural], xviii, 4, 13, 21; [who] "arrive [are born] with a predisposition for helpfulness and cooperation," 43; [who] "come to culture ready to be helpful," 44; "[who] are altruistic by nature," 47; and "[who] are social in many ways," 166.

270 Nancy Abrams, *A God That Could Be Real*, 136.

271 Bruce Alexander, *The Globalization of Addiction*, 2008, 86-87.

272 Samuel Bowles et al, *A Cooperative Species: Human Reciprocity and its Evolution*, (Princeton, NJ: Princeton University Press, 2011), 16.

273 Matt Ridley, *The Origins of Virtue*, 1996, 5.

On Evolution

cooperate to compete, but more to cooperate and collaborate with others "who may be complete strangers." This he terms "hyperprosociality... not a learned tendency but a genetically coded trait found only in *H. sapiens*..."[274] This is completely at odds with what Darwin propounded, the near exclusive emphasis on competition in natural selection. [275]

Ridley explained the intimate relationship between genetic nature and learned nurture,[276] concluding that "Nature versus nurture is dead. Long live nature via nurture."[277] But he does not enter the combative cooperation versus competition arena - there are no references to either in the book's index - leaving us to decide for ourselves which is paramount. However, his silence here may indicate that, for him, cooperation and competition are both innate, then learned more about after birth. This seems reasonable.

Going on to show us the way to the success of our species in the face of the great challenges of the day, including disease, climate change, global hunger, mistakes (e.g., many mutations), and poverty, Nowak et al, say that we cannot reach utopia.[278]

I disagree. Utopia is always possible: we have only to do a series of right things. First Nations Grandfather Rolling Thunder said, "...we can heal the Earth and make it like it was a long time ago."[279] An "earthly paradise" is possible, says Wright, if we "shape it, share it, and look after it."[280]

---

274 Curtis Marean, "The Most Invasive Species of All," *Scientific American*, August 2015, Vol. 313, No. 2, 35. It is open to question whether this trait is exclusive to human beings.

275 Kelly Clancy has written a magazine article, titling it "Survival of the Friendliest." (New York: Nautilus, Issue 046, March 23, 2017).

276 Matt Ridley, *The Agile Gene: How Nature Turns on Nurture* (Toronto: HarperPerennialCollins, 2003).

277 Ibid, 280.

278 Martin Nowak et al, *SuperCooperators*, 282.

279 Karen Speerstra, *The Green Devotional: Active Prayers for a Healthy Planet* (San Francisco: Conari Press, 2010), 244.

280 Ronald Wright, *A Short History of Progress*, 9.

Was Darwin racist?

Charles De Paolo's book, *The Ethnography of Charles Darwin: A Study of His Writings on Aboriginal Peoples*, is a trailblazing "review of the scholarship" on Darwin's writings about Indigenes, yeoman work bearing a treasure of information.[281]

I had previously suspected that Darwin was racist, and De Paolo's and others' work confirmed, unfortunately, that this was so. Darwin held the Fuegian peoples of Tierra del Fuego in particular disdain, and his dehumanizing of them became an attempt to provide a living "missing link" between apes and humans. De Paolo records examples of Darwin's racism, here listed in point form; Darwin:

- "positioned the Fuegians and other living savages in a chain which ran from ape to European."[282]
- felt that "The Alacaluf [one of the Fuegian tribes] were not human beings at all."[283]
- "went so far as to dislodge the people of the Archipelago [i.e., the Fuegians], zoologically, from the human genus."[284]
- felt that "Fuegian nakedness, painted bodies, tangled hair, and discordant voices...in his mind, confirmed their sub-humanness... [Darwin] referred to them...as 'unbroken savages' and as 'gifted animals' - as creatures somewhere between man and beast."[285]
- drew "a closer affinity between savages and higher primates than between the former and Western man."[286]
- wrote, "[N]ot understanding language of Fuegian puts [them] on par with Monkeys."[287]

---

281 Charles De Paolo, *The Ethnography of Charles Darwin: A Study of His Writings on Aboriginal Peoples*, (Jefferson, NC: McFarland & Co., Inc., 2010).
282 Ibid, 85.
283 Ibid, 135.
284 Ibid, 136.
285 Ibid
286 Ibid
287 Ibid, 143.

On Evolution

Browne says that, for Darwin, absurdly, "...a single glance at the respective facial expressions of [a Maori and a Tahitian] convinced him one is a savage, and the other a civilized man."[288]

To support his ideas about a competitive natural selection that would eliminate the "unfit," Darwin referred in *The Descent of Man* to "the immorality of savages"[289] and, as well, "the imbecile, maimed, and other useless members of society,"[290] all unfit for survival. Further, he said it was his intention to "trace a perfect gradation from the mind of an utter idiot, lower than that of an animal low in the scale, to that of a Newton."[291]

Such ideas continue in Darwin's autobiography where he supposed that "[the] presence of much suffering [in English society] agrees with the theory of natural selection [via competition]."[292] Unable or unwilling to see outside the box he had created for himself, he did not recognize that the presence of much suffering in society might equally have agreed with its lack of mutual aid. This represents a serious bias in his thinking, one that is difficult to reconcile with his image as a pre-eminent scientist.

Yet it is consistent with his Malthusian views of the poor, hungry, disabled, and "savage," all as "unfit" in the struggle for existence. Darwin professed to being "ignorant of history, politics, and moral philosophy,"[293] but Desmond et al say that his notebooks made it clear that competition, sexual inequality, imperialism, and racial extermination were part of his agenda before he embarked on the voyage of the Beagle.[294]

Browne says that Darwin first felt some sentiment toward the First Nations population of Patagonia during the time that he spent there. However, after he saw the retaliatory damage done by First Nations

---

288 Janet Browne, *Charles Darwin: Voyaging*, 312.
289 Charles Darwin, *The Descent of Man*, 116.
290 Ibid, 121.
291 Ibid, 123.
292 Charles Darwin, *The Autobiography of Charles Darwin*, edited by Nora Barlow. (London, UK: Collins, 1958), 90.
293 Ibid, 55.
294 Adrien Desmond et al, 1991, xxi.

people to Spanish estates he encountered along the way, he changed his mind, saying: "It is clear to me that [Spanish General] Rosas ultimately must be absolute Dictator...of this country."[295]

Desmond et al wonder why, "...given [his] family's emotional anti-slavery views..., Darwin's biologizing of genocide should appear to be so dispassionate."[296] Though Darwin was apparently opposed to slavery himself, he supported the Spanish war against First Nations peoples in South America, and the expropriation of their land. According to Bowlby, Darwin said:

> This war of extermination, although carried on with shocking barbarity, will certainly produce great benefits, it will at once throw open four or five hundred miles in length of fine country for the produce of cattle.[297]

Darwin merely resigned himself to this slaughter.[298] In fact, say Desmond et al, Darwin agreed with the elimination of the "lower" races by the "higher" races[299] His "mechanism...was modern-day massacre."[300] "Extermination [is] an axiom of nature," said Darwin, "[w]ith Nature moving forward, crushing skulls underfoot."[301]

Montagu said that Darwin eventually went so far as to predict that, at "some future period, not very distant as measured by centuries, the civilized races of man will almost certainly exterminate . . . the savage races throughout the world."[302] Darwin, according to Himmelfarb,

---

295 Janet Browne, *Charles Darwin: Voyaging*, 255. Rosas did become a dictator, ruling Argentina until 1852 before fleeing to England and his death in exile 20 years later.
296 Adrien Desmond et al, *Darwin's Sacred Cause*, 2009, 150.
297 John Bowlby, *Charles Darwin: A Biography*, (London: Hutchinson, 1990), 181.
298 Adrien Desmond et al, 2009, 146.
299 Ibid, 318.
300 Ibid, 149.
301 Ibid, 151.
302 Ashley Montagu, *Darwin: Competition and Cooperation*, 108.

On Evolution

"was not averse to the idea that some races were more fit than others . . ."[303] "By biologizing colonial eradication," Desmond et al write, "Darwin was making 'racial' extinction an inevitable evolutionary consequence."[304]

Entertaining notions of helping out in this mission, and convinced that he might face mortal danger during his voyage, Darwin bought pistols before he left,[305] intended to "keep the natives...quiet."[306] Desmond et al further quote Darwin as saying, "We shall have plenty of fighting with those d_____ Cannibals...it would be something to shoot the King of the Cannibal Islands."[307]

It is peculiar that Darwin is so often described as merely genial and timid. Fond of shooting animals into middle age, he participated in a massacre of flightless birds during the journey with FitzRoy.[308] When exploring the Galapagos Islands, he said that he'd thrown a lizard as far as he could, cruelly - to watch it struggle.[309] While Desmond et al describe Darwin as a "gentle naturalist" and "the most gentlemanly gentleman anyone had ever met,"[310] Quammen counters, saying, "Darwin was a selfish and ruthless man in some ways . . . selfish and ruthless mainly in service to his work."[311] Like Kropotkin, Darwin was probably as complex a man as any, with a full range of qualities and faults.

---

303 Gertrude Himmelfarb, *Darwin and the Darwinian Revolution*, 416.
304 Adrien Desmond et al, *Darwin's Sacred Cause*, 2009, 149.
305 John Bowlby, 1990, 112.
306 Adrien Desmond et al, *Darwin*, 1991, 104.
307 Ibid. Here Darwin expresses an almost adolescent view of Indigenes.
308 Janet Browne, *Charles Darwin: Voyaging*, 204.
309 Robert Jastrow, ed., *The Essential Darwin*, 40.
310 Adrien Desmond et al, 2009, xvi.
311 David Quammen, *The Reluctant Mr. Darwin: An Intimate Portrait of Charles Darwin and the Making of His Theory of Evolution*, (New York: W.W. Norton & Company, Inc., 2006), 236.

James G. Duncan

Variability and violence in *The Origin of Species*

While key aspects of Darwin's logic regarding cooperation, mutual aid, and sociability in natural selection remained from the first to the last editions of *The Origin of Species*, other variations in the text were common. Morse Peckham examined the six editions published from 1859 to 1872, and observed that there were many changes made. Says Peckham, "Of the 3878 sentences in the first edition, nearly 3000, about 75%, were rewritten from one to five times . . .over 1500 sentences were added, and . . .nearly 325 were dropped."[312] Was this a sign that Darwin lacked confidence in his research and writing? One wonders.

One significant change from the first edition of *The Origin of Species* is reflected in conceptions of power. The final two sentences of the first edition recognized only the power of evolution; Darwin wrote:

> Thus, from the war of nature, from famine and death, the most exalted object which we are capable of conceiving, namely, the production of the higher animals, directly follows. There is grandeur in this view of life, with its several powers, having been originally breathed into a few forms or into one; and that, whilst this planet has gone cycling on according to the fixed law of gravity, from so simple a beginning endless forms most beautiful and most wonderful have been and are being evolved.[313]

However, the final edition of the *Origin* included the power of the Creator; it read:

> Thus, from the war of nature, from famine and death, the most exalted object which we are capable of conceiving, namely, the production of the higher animals, directly follows. There is grandeur in this view of life, with its several powers, having been

---

312 Morse Peckham, *"The Origin of Species" by Charles Darwin: A Varorium Text*, (Philadelphia: University of Philadelphia Press, 1959), 9.

313 Charles Darwin, *The Origin of Species*, 1968, 1st edition 1859, 459-60.

On Evolution

originally breathed *by the Creator* into a few forms or into one; and that, whilst this planet has gone cycling on according to the fixed law of gravity, from so simple a beginning endless forms most beautiful and most wonderful have been and are being evolved.[314]

The second sentence in this quotation is poetic and widely cited, perhaps inducing a warm feeling toward Darwin as evolutionary father.

But the first sentence is not warm, nor is it commonly quoted. But I must quote it again: "Thus, from the war of nature, from famine and death, the most exalted object we are capable of conceiving, the production of the higher animals, directly follows."

What Darwin is saying, in the final words of *The Origin*'s last edition, is that the "higher animals" have evolved from "the war of nature, from famine and death." Taken together with the sentiment in the second sentence, Darwin's view of the evolution of the "higher animals" takes on a decidedly macabre and Malthusian cast. Too, the God of his understanding evidently had no room for love and compassion. As noted for Kropotkin, Darwin did not have a place for a benevolent Higher Power in his evolutionary ideas.

In fact, Darwin was quite forward with violence in the first edition of *The Origin of Species*. For instance, this includes "heavy destruction,"[315] "tremendous destruction,"[316] "severe struggle,"[317] "great destruction,"[318] "battle of life,"[319] "war,"[320] "war of nature,"[321] "very severe competition"[322]

---

314 Charles Darwin, *The Origin of Species*, 6th edition 1872, 463. Italics mine.
315 Charles Darwin, *The Origin of Species*, 1968, 1st edition 1859, 119.
316 Ibid, 121.
317 Ibid
318 Ibid, 129.
319 Ibid, 127, 130, 170.
320 Ibid, 126, 459.
321 Ibid, 128, 459.
322 Ibid, 152.

and the following sentence, removed in later editions, describing nothing but violence in Nature:

> The face of Nature may be compared to a yielding surface, with ten thousand sharp wedges packed closely together and driven inwards by incessant blows, sometimes one wedge being struck, and then another with greater force.[323]

The Origin of Species appeared in six editions during Darwin's lifetime, the last in 1872. However, the book and its revisions were only an abstract of a larger work that Darwin planned to write, but never did. In an abstract, he had no obligation to cite his sources. This deficiency was never corrected and it, too, is rarely cited.

HPMA, Evolution, and the Future: Brief Notes

Murray Bookchin provides some details of an extensive updating and deepening of Kropotkin's ideas in *Mutual Aid*, referring to the examples of Trager (1970) and Margulis (1981).[324] Cohen notes that mutualism - mutual aid - is largely neglected in modern studies of ecology,[325] and Sapp gives much credit to Kropotkin as a challenge to neo-Darwinism.[326] Morris states, however, that many of today's evolutionary theorists seem not to have even heard of the Russian prince.[327]

In the minds of some, Kropotkinist ideas about mutual aid in natural selection have been left far behind. However, the real spirit of

---

323 Ibid, 119.

324 Murray Bookchin, *The Ecology of Freedom: The Emergence and Dissolution of Hierarchy*, revised edition (Montreal, QC: Black Rose Books, 1991), 358–61.

325 Hiroya Kawanabe et al, eds., *Mutualism and Community Organization: Behavioral, Theoretical, and Food-Web Approaches*, (Oxford: Oxford University Press, 1993), 414.

326 Jan Sapp, *Evolution by Association: A History of Symbiosis*, (Oxford: Oxford University Press, 1994, 20-23).

327 Brian Morris, *Kropotkin: The Politics of Community*, 2004, 144, 148.

Kropotkinism has not been lost. Fritjof Capra almost takes words out of the prince's mouth, saying in 1998, "Life is much less a competitive struggle for survival than a triumph of cooperation..."[328] He adds what Kropotkin was saying in the late 1800s and early 1900s, namely that:

> [W]e are now beginning to see continual cooperation and mutual dependence among all life forms as central aspects of evolution.[329]

Darwinian apologists still abound, virtually falling all over themselves to praise him and his work, and uncritical acceptance of his thinking remains. Van der Dennen repeats Darwin's notion that social instincts have been acquired by natural selection,[330] Carroll extols Darwin's virtues before presenting *The Origin's* first edition,[331] while Palmer, like Darwin, raises the spectre of Malthusianism,[332] even now when ecology is being increasingly linked with sustainability.

Perhaps swept up by Darwin's popular but terribly flawed theory of natural selection by competition, Thayer puts forward the notion that it provides an explanation for warfare and ethnic conflict.[333] J.W. Burrow observes that one military man said war was "a biological necessity."[334] Such arguments hold no water whatsoever

---

328 David Loye, ed., *The Evolutionary Outrider: The Impact of the Human Agent on Evolution*, (Westport, CT: Praeger, 1998), 46.

329 Ibid

330 Johan van der Dennen et al, eds., *The Darwinian Heritage and Sociobiology*, (Westport, CT: Praeger, 1999), 277.

331 Joseph Carroll, ed., *On "The Origin of Species": Charles Darwin*, (Orchard Park, NY: Broadview, 2003).

332 Trevor Palmer, *Perilous Planet Earth: Catastrophe and Catastrophism through the Ages*, (Cambridge: Cambridge University Press, 2003).

333 Bradley Thayer, *Darwin and International Relations: On the Evolutionary Origins of War and Ethnic Conflict*, (Lexington, KY: University Press of Kentucky, 2004).

334 Charles Darwin, *The Origin of Species*, 1968, 1st edition published 1859, 45.

when cooperation, mutual aid, and sociability are factored into the equation.[335]

Others fail to see Darwin's faulty conclusion that competition is all in natural selection.[336] Darwin was a relentlessly persuasive force in putting forth competition as the sole engine for natural selection, with mutual aid not even on his radar. It must be said that he needed no one - not even his willing followers - to add to what he said in *The Origin* and *The Descent of Man*. His theory was wantonly competitive from the start.

Miriam Simos initially cites Darwin's emphasis on competition as valid, but proceeds to ambivalence about the matter, saying there is a balance between competition and cooperation.[337] She goes on to say that cooperation is the main factor of evolution.[338]

Karen Armstrong is also persuaded that Darwin's emphasis on competition is sound, and that cooperation and compassion for others are not. She says, "…in the deepest recess of their minds, men and women are indeed ruthlessly selfish,"[339] ignoring evidence for the cooperation and sociability that are innate in all of God's creatures. Perplexingly, Armstrong also observes of Indigenous peoples that "The clan would

---

335 While the Indigenous peoples of the Americas did have conflicts, outright warfare was unknown before the advent of white imperialism. In this context, we can refer to words from Lakota warrior, Crazy Horse, "[W]ith the white warriors it was killing every day, killing all the time." (Mari Sandoz, *Crazy Horse*, [Lincoln, NE: University of Nebraska Press, 1961], 315). Similarly, holy man Black Elk said whites "killed and killed [bison] because they liked to do that." (Black Elk, *Black Elk Speaks*, as told through John G. Neihardt, [Lincoln, NE: University of Nebraska Press, 1979], 213).

336 See, for example, Daniel Dennett, *Darwin's Dangerous Idea: Evolution and the Meanings of Life*, (New York: Simon & Schuster, 1995).

337 Miriam Simos, *The Earth Path: Grounding Your Spirit in the Rhythms of Nature*, (San Francisco: Harper Collins Publishers, 2004, 42-43).

338 Ibid, 47-49.

339 Karen Armstrong, *Twelve Steps to a Compassionate Life*, (Toronto: Alfred A. Knopf Canada, 2010), 13.

On Evolution

survive only if members subordinated their personal desires to the requirements of the group . . ."[340]

Armstrong blames this selfishness on the old reptilian brain, as if reptiles are not capable of love and compassion. Crocodile mothers carry their young in fearsome, toothsome mouths. I am reminded of a story about birds, passed along by a friend. Birds, it has been shown, are reptilian. Called "God's Wings," it is paraphrased thus:

> After a fire in Yellowstone Park, forest rangers climbed up a mountain to determine what damage the fire had caused. At one point, a ranger found a quail "literally petrified in ashes." After it was knocked over, three tiny chicks fled from under their dead mother's wings. The mother bird had gathered them together under her to protect them, knowing that the fire would kill her babies. The heat from the fire burned her to death, because she had been willing to die, that those under her wings might live.

An Al-Anon friend shared her reaction to this story, saying, "The power of the maternal parallels the power of nature and the power greater than ourselves."

The Oxford English Dictionary has a subtle Darwinian bias. The word competition has an ecological definition therein, while cooperation does not have even a biological definition. As well, the dictionary cites mutualism as a biological word meaning, "A condition of symbiosis in which two associated organisms contribute mutually to the well-being of each other." Though mutual aid is a far better descriptor of cooperation than mutualism, it is not a dictionary term.

Darwin emphasized natural selection through individual competitive struggle. This was also stressed by his followers and the role of innate cooperation in natural selection became underestimated and undermined. Kropotkinist ideas about mutual aid - countering Darwin's competition and the "struggle for existence" - were lost. Evidence of

---

[340] Ibid, 30.

innate sociability and the importance of the common welfare were ignored in favour of ideas supporting the competitive acquisition of power and possessions.

The planet-wide reach of effective Twelve Step fellowships, and the successes of First Nations societies, is powerful evidence that evolution comes about chiefly through the Higher Powered variant of Kropotkinist mutual aid. This is a matter of considerable import for humanity, with deep, broad implications.

It is clearly the case that Darwin's ideas regarding the competitive "struggle for existence" find no place in the success of Twelve Step societies, and more lately, First Nations societies now allowed, somewhat, to prosper. Rather, their development and survival has always depended upon mutual aid, the individual freedom within the common welfare described by Kropotkin, and, guided by Higher Powers, practiced successfully by the Twelve Step founders and by ancient First Nations.

Many did not take Kropotkin's ideas seriously, but those Twelve Step pioneers did in laying down the foundations of Twelve Step HPMA for A.A., Al-Anon, and other Twelve Step societies. They placed Kropotkin's work alongside that of Carl Jung, William James, and others, to strengthen that base.

"In giving to the individual member," Katz et al write, fittingly, "[mutual aid] groups," such as A.A., "have thus also given to the larger society."[341] Furthermore, what I believe to be the promising case, they add that mutual aid groups can be "a vehicle for *social* change, of an untapped and even unimagined potency."[342] Humanity could take some valuable lessons from this if it hopes to survive.

The objectives of the next two chapters are to examine authority, spirituality, serenity, and leadership in age-old First Nations societies; to raise awareness of parallels between First Nations HPMA and younger Twelve Step; and to look at ideas for the wider application of Higher

---

341 Alfred Katz et al, *The Strength in Us*, 240.
342 Ibid, 241.

On Evolution

Powered Mutual Aid and its creative, democratic, ecological, liberating, moral, peaceful, self-responsible, and spiritual principles. This is partly in order to illustrate that the principles of Twelve Step HPMA are, as the Twelve Step pioneers said more than once, ancient and universal. First Nations may have much to teach their younger Twelve Step siblings.

CHAPTER 4

First Nations and Twelve Step Societies

Go to the ant; consider her ways, and be wise:
which having no guide, overseer,
or ruler, provides her meat in the summer, and
gathers her food in the harvest.

PROVERBS 6:6-8

THE CULT OF individualism at the expense of community is at an end. Competitive individualism, though not as evident as cooperation and mutual aid, has attempted to operate outside supporting relationships and the common welfare. This has proven to be a destructive force for the past 3,000 years, particularly the last 500. Unbounded individualism is not healthy for individuals, families, communities, countries, or for the world, since it denies the natural unity of the individual and the common welfare. The single-minded competitive pursuit of the individual good has brought, not the common welfare, but the ongoing destruction of the Earth itself. We must do better.

It had to end eventually. Economic, environmental, and social collapse may be inevitable now, unless humanity "remembers" to be ecological. Garret Hardin's infamous "The Tragedy of the Commons"[343] described irresponsible individual behaviour, overgrazing one's animals on land held in common with others. Such irresponsible behaviour by

---

343 Garret Hardin, "The Tragedy of the Commons." *Science: Journal of the American Association for the Advancement of Science* 162, no. 3859 (December 13, 1968).

an individual, and nations of individuals, worldwide, and resulting ill effects, could have been avoided by self-regulating, self-responsible individuals who paid heed to the common welfare and the long-term maintenance of the commons.

I have put forward the idea that the Twelve Step movement can help us end our progress on the devastating path upon which we stumble, often under the influence of alcohol and other drugs. But we can stop on the eve of destruction to achieve a balance of cooperation and competition within structures of Higher Powered Mutual Aid. This is a suggestion for a return to HPMA as the foundation for democratic, ecological, moral, social, and spiritual reorganization and renewal, locally and globally.

Beyond Kropotkin's secular mutual aid and the benign anarchy of Twelve Step HPMA, are the far older spiritual and healing traditions of First Nations, traditions that Darwin, Kropotkin, and others discounted, no doubt on "scientific" grounds. Consideration of some key aspects of First Nations organization in the Americas, for example, show that there are strong similarities between First Nations and Twelve Step Higher Powered Mutual Aid. Tentative conclusions indicate that, together, the two types of society might provide a wealth of applicable ideas and methods to bring about the rescue of endangered species such as ours.

Bill W. referred to A.A. as "spiritual kindergarten" in a 1954 letter;[344] one A.A. speaker echoed this, describing A.A. as the "kindergarten to good living." As noted, Bill W. said that the principles of Twelve Step HPMA recovery are all based on "ancient sources."[345] How ancient is ancient? In my view, the principles of Higher Powered Mutual Aid are universal and ecological. They are the very stuff of healthy life itself, innate in all Creation from the start, existing until the sun shines no more.

Mainstream societies have not used Higher Powered Mutual Aid in social organization for a long time, and are increasingly beset by ecological, economic, mental, moral, physical, social, and spiritual illnesses. Our

---

344 Alcoholics Anonymous, *As Bill Sees It: The A.A. Way of Life*, 95.
345 *Alcoholics Anonymous Comes of Age*, 231; *The Language of the Heart*, 345.

behaviour and actions have brought many negative consequences: in time worn parlance, the chickens have come home to roost. However, HPMA has already been relearned in Twelve Step fashion, and greater success can come about by examining its roots and applying its time-proven principles.

Though most of my own experience with HPMA has been gained in Twelve Step societies, especially Al-Anon, I have also come to some understanding of First Nations traditions through personal experience, and through studying and reading about their societies and their early contacts with Europeans. Attuned to the principles of Twelve Step Higher Powered Mutual Aid, it gradually dawned on me that there were strong parallels between Twelve Step and earlier First Nations societies.[346]

In fact, it appeared that these parallels were very strong, and there came a convergence in my understanding of how things worked in the two kinds of society. Taken together with studies in social ecology, including the works of Peter Kropotkin and Murray Bookchin, I began to think, partly through intuition, that the similarities between Twelve Step group and First Nations circle societies might represent something important. Inspired, I carried on with the research and writing for this book.

In this chapter, I first document information regarding First Nations circle societies at the time of European contact in the early 1500s; these are described in more detail by Graveline[347] and Sioui.[348] The focus will turn in the next chapter to some specific parallels between older First Nations societies, and those of the first two fellowships of the Twelve Step movement, A.A. and Al-Anon. It should be noted that descriptions and

---

346 I name most of the Indigenous peoples of North, Central, and South America, "First Nations," a Canadian innovation that avoids use of the word "Indian." Métis and Inuit peoples are not included in this book; they have their own valuable traditions which I cannot cover here due to lack of space.

347 F.J. Graveline, *Circle Works: Transforming Eurocentric Consciousness*, (Halifax, Nova Scotia: Fernwood Publishing, 1998).

348 Georges E. Sioui, *Huron-Wendat: The Heritage of the Circle*, revised edition, (Vancouver, BC: University of British Columbia Press, 1999).

On Evolution

observations herein, about First Nations societies of the Americas, are only examples, scratching the surface and in outline at that.

Despite the deserved condemnation of the overall manner in which Europeans presented themselves in the home of First Nations peoples, mutual aid and sharing between them did take place in some cases.[349] This makes a lot of sense; we are a complex, creative, and sometimes generous species. Perhaps we are like most of life, wanting to use our complexity and creativity wisely, when given half a chance, even when creative power is distorted by anti-ecological bias, hierarchical authoritarianism, and resulting dysfunction.

It is an irony of history that the Twelve Step pioneers went so far afield to strengthen the theoretical base for the effective use of Higher Powered Mutual Aid. They found, or shored up, many ideas about secular mutual aid in the works of a Russian prince rather than in the more comparable First Nations HPMA. Said one Mohawk Elder involved in First Nations substance abuse recovery work:

> One of our main teachings is that the 12 Steps are very compatible with the culture-based recovery program of the Code of Handsome Lake, a [First Nations] temperance movement that began in the mid-1700s.[350]

However, like Kropotkin, the Twelve Step pioneers made use of the written word, not only the verbal traditions used by First Nations peoples.

---

349 See Peter Nabokov, *Native American Testimony: A Chronicle of Indian–White Relations from Prophecy to the Present, 1492–1992*, (New York: Penguin Books, 1991, 69, 72); Jack Weatherford, *Savages and Civilization: Who Will Survive?* (New York: Crown Publishers, Inc., 1994, 10); Peter Kulchyski et al, eds., *In the Words of Elders: Aboriginal Cultures in Transition*, (Toronto, ON: University of Toronto Press, 2003, 40).

350 Personal communication. Paula Gunn Allen is not persuaded that the Code of Handsome Lake was beneficial for female Iroquois Elders, traditional tribal leaders. (Paula Gunn Allen, *The Sacred Hoop: Recovering the Feminine in American Indian Traditions*, [Boston: Beacon Press, 1986], 32-33).

Higher Powered Mutual Aid was once the bedrock of European cultures. Simos notes that, "...the origins of European civilization... lay in cultures that honored the earth and valued cooperation over ruthless competition and war."[351] But by the time First Nations societies were encountered in the Americas, the Europeans had forgotten HPMA, and were perhaps unable to any longer recognize its purposes and functions in social organization. Habitual warfare in hierarchical authoritarian societies, accompanied by trauma and medicated with alcohol, might accomplish that. Resulting serious diseases drew the Europeans from their homes for greener pastures in the Americas and beyond.

Over the past 500 years, the green pastures of the Americas have become as brown as those of old Europe. The spiritual diseases stemming from authoritarianism, fear, insecurity, possessiveness, and unbounded individualism, have taken their toll. Nabokov recounts the words of one Elder, who said of whites, "The love of possessions is a disease among them."[352] Whites' greatest objective, said another, was ". . . to acquire possessions - to be rich. They desire to possess the whole world."[353]

By comparison, First Nations peoples possessed only what sustained them from day to day, month to month, or season to season, giving them an environmental impact, an "ecological footprint," much smaller than that of the Europeans.[354] First Nations possessions were minimal, partly because some were nomadic, following, for example, the immense bison herds that they depended upon. As well, game was so plentiful before the

---

351 Miriam Simos, *The Earth Path*, 22.

352 Peter Nabokov, *Native American Testimony*, xviii.

353 Ibid, 22. Carl Jung cited the greed of the West. (Meredith Sabini, ed., *The Earth has a Soul: C.G. Jung on Nature, Technology & Modern Life*, [Berkeley, CA: North Atlantic Books, 2008], 125).

354 For more explication, see Mathis Wackernagel et al, *Our Ecological Footprint: Reducing Human Impact on the Earth*, (Gabriola Island, BC: New Society Publishers, 1996).

On Evolution

coming of the white man that First Nations peoples had no incentive to work for a living.

First Nations Lakota warrior Crazy Horse had this to say:

> We did not ask you white men to come here. The Great Spirit gave us this country as a home. You had yours. We did not interfere with you. The Great Spirit gave us plenty of land to live on, and buffalo, deer, antelope and other game. But you have come here; you are taking my land from me; you are killing off our game, so it is hard for us to live. Now, you tell us to work for a living, but the Great Spirit did not make us to work, but to live by hunting. You white men can work if you want to. We do not interfere with you, and again you say why do you not become civilized? We do not want your civilization! We would live as our fathers did, and their fathers before them.[355]

In the following sections, I discuss European perceptions of, and influences upon, First Nations societies, and the interwoven subjects of authority, fear, leadership, liberty, serenity, and spirituality. These aspects of First Nations cultures are of direct relevance to Twelve Step societies. The focus here is on European contacts with First Nations societies at the southernmost part of the Americas, some brief, and later of longer duration. Portugal's Ferdinand Magellan and England's Francis Drake passed through there first in the early 1500s, and in 1831, the English captain Robert FitzRoy was accompanied aboard the *Beagle* by a budding naturalist, Charles Darwin.

Authority in First Nations Societies

The placement of authority in First Nations' varieties of a Higher Power, a Creator or Supreme Authority, is central to First Nations societies, as it is central to Twelve Step fellowships. The subject of authority in First

---
[355] Peter Matthiessen, *In the Spirit of Crazy Horse*, (New York: Viking Penguin, 1991), ix.

James G. Duncan

Nations cultures is complex and varied, constituting a subject worthy of separate investigation. One contact indicated that, in Canada alone, there remain over 600 different First Nations, each with their own circle traditions.[356] This is consistent with the ecological diversity represented around the Americas and elsewhere.[357] Kulchyski et al note that "[The] power of the [First Nations] world always works in circles;"[358] there is no hierarchy in the circle.

Wright portrays the cultures of five different First Nations societies of the Western Hemisphere, the Inca of the Andes, the Aztec and Maya of northern Central America and Mexico, and the Cherokee and Iroquois of southeastern North America. He comments on the exercise of human authority among the Aztecs,[359] describing them as "callow imperialists" and a warrior culture, concluding that they and the invading Spanish "deserved each other."[360] Said Wright also, Aztec society was a "loose pyramid of city-states" whose leader was the "Speaker." This suggests confederacy as well as hierarchy, but Wright does not expand nor does he explain the term Speaker.

In describing the interweaving of spirituality and authority among the Aztecs, Wright seems skeptical at times,[361] while at others he is more reverential.[362] He makes further comments intertwining environmental change, human authority, and war, in spirituality,[363] giving one a sense of the sophistication of First Nations cultures. However, one does not come away with a good understanding of Aztec spirituality.

---

356 Personal communication.

357 North America is known by some First Nations as "Turtle Island." In geology's "plate tectonics," the North American plate is moving west-southwest at about 2.5 centimetres per year. Turtles move slowly, too.

358 Peter Kulchyski et al, eds., *In the Words of Elders*, 2003, xix.

359 Ronald Wright, *Stolen Continents*, (New York: Viking, 1992), 15–47.

360 Ibid, 31.

361 Ibid, 19–21.

362 Ibid, 32–34.

363 Ibid, 34–36.

On Evolution

Wright says that the Aztec especially, and the Maya and Inca, developed forms of hierarchy with some human authority, held in a class of people with better or even exclusive access to Higher Powers. Tremendous powers might have accrued to those who had such access to Higher Authority, or who were seen to have such access. Bookchin provides one explanation for the rise of priestly classes in *The Ecology of Freedom*.[364]

In depicting the anarchic/democratic structures of the First Nations Cherokee federation[365] and the Iroquois Confederacy,[366] Wright provides some details of the place of Higher Powers in First Nations that, put simply, promote individual liberty within the common welfare. Said Wright, the Iroquois Confederacy served as a model for the U.S. Constitution, and for some of the ideas of Marx and Engels.[367]

However, one of Wright's informants said that neither the U.S. nor the U.S.S.R. had been able to use the Iroquois model properly,[368] and it may be that neither really understood it. Old habits and policies, rooted in faraway, overseas struggles, were not sufficiently malleable to allow for the re-application of the much older ideas of First Nations peoples. Thus, the anarchic female-centric democracy of the Iroquois Confederacy, rooted in the natural environment, may simply have been translated into male-centric hierarchical, authoritarian government with its implementing bureaucracy.

The smaller societies of First Nations such as the Cherokee and Iroquois, said Wright, recognized the authority of Higher Powers in beneficial, benign, and effective anarchies, with little if any hierarchical human authority. "Beneficial," "benign," and "effective" are used here to describe First Nations democracies using HPMA, since they promote individual freedom within the common welfare, without human authority.

---

364 Murray Bookchin, *The Ecology of Freedom: The Emergence and Dissolution of Hierarchy*, (Montreal, QC: Black Rose Books, 1991), 89-118.
365 Ronald Wright, *Stolen Continents*, 15–47.
366 Ibid, 118–21.
367 Ibid, 116-17. For more detail see Paula Gunn Allen, *The Sacred Hoop*, 218-20.
368 Ibid, 118.

The most important similarity between Twelve Step organizations and those of the Cherokee and Iroquois, is the central place of Higher Authority. In each of these cultures, authority resides in various forms of a Higher Power, not in any one person or class of persons. As well, in each of these societies, ancient First Nations and the more recent Twelve Step, the common welfare comes first, giving individuals a safe, secure freedom so that each may define and carry out their creative potential and purpose.

Nabokov says that the head of the U.S. Bureau of Indian Affairs from 1933 to 1945, John Collier, thought the Taos First Nation was "a model for all mankind, a '"Red Atlantis"'...a community that successfully integrated the needs of the individual with those of the group."[369] On the same page, Nabokov quotes Collier as saying, "[First Nations] had what the world has lost. They have it now. What the world has lost, the world must have again lest it die...the ancient reverence and passion for the earth and the web of life." Twelve Step HPMA societies reach for such heights, too, as we will see, having a benign anarchy of democratic groups that are similarly effective in promoting individual healing and freedom within the common welfare, with local and global potential.

The key effect of individual liberty within the common welfare, guided by Higher Authority, is peace or serenity, the "priceless gift" so termed by Bill W.[370] Many observations show that First Nations societies exhibited this peace and serenity when the Europeans arrived on the shores of the "New World." The following constitutes only a small sample of the available material on the subject of serenity in First Nations cultures.

". . . all virtue in a certain gentleness . . ."

Many early European descriptors of First Nations societies included "benevolent," "gentle," "kindly," and "shy," on the one hand, or "apathetic"

---

369 Peter Nabokov, *Native American Testimony*, 305.
370 Alcoholics Anonymous, *Twelve Steps and Twelve Traditions*, 74.

On Evolution

and "lazy" on the other. While thinking about the similarities between First Nations and Twelve Step Higher Powered Mutual Aid, I wondered if these latter words might have been synonyms or euphemisms for the serenity that Europeans had forgotten or, when encountered, misconstrued. If that was so, it may have been likely that they could not recognize serenity, or its value, even when it was so widely found in First Nations societies of the time.

If First Nations peoples had lived in HPMA societies for thousands of years, then, as Twelve Steppers might recognize, they could well have been mostly gentle and serene peoples, not generally given to extremes of human emotions, and therefore better able to coexist in harmony. Perplexed by First Nations serenity, the Europeans seem to have been equally frustrated by First Nations who did not have a "chief," an English word,[371] to exhibit the human authority with which they were familiar.

A problem for some modern First Nations is the dominance of such "chiefs," an emulation of authoritarian hierarchically-organized whites who wanted, and still want, to speak to elected leaders. One First Nations Elder said ". . . what's happening now is the [elected] Chiefs have all the say . . . not the community, no more, just Chiefs."[372] As we will see, Peter Kropotkin admitted that the small communities of his time failed due to this election of leaders.

In Patagonia and Tierra del Fuego of southern South America, the earliest contacts with First Nations were made by Magellan and Drake, typical ultra-authoritarian ship captains. Accounts of their explorations in the early 1500s, before the large-scale European invasion of the Americas, make it seem that First Nations people existed in more harmony than many have imagined.

McKew Parr says that Magellan, the first European explorer in the area in 1520, saw fires at night but "In the day time, he could see no

---

371 Personal communication.
372 Peter Kulchyski et al, eds., *In the Words of Elders*, 2003, 339.

traces of the shy natives."[373] During Drake's voyage 20 years later, contact was made with Patagonians south of the river Plate, north of Tierra del Fuego. They were seen as "good kindly folk" and, to Drake's chaplain, were "kinder to him than many of his clerical brethren at home."[374]

After the capture of three Fuegians on an earlier voyage, captain FitzRoy was surprised by their apparent lack of curiosity about their surroundings in England. To him, this marked them as "intellectually idle."[375] Hazlewood feels that the captives may not have been used to the warmer climes to the north, or that they might have been disconcerted, dumbstruck, fearful, or numbed. This may be partly correct.

But what FitzRoy found so curious, even contemptible, might simply have been related to serenity-creating First Nations societies. It may be that FitzRoy's three captives remained relatively serene and unimpressed by outward shows of culture.[376] From his European perspective, FitzRoy failed to recognize how First Nations people centred knowledge, power, and value within their Creator, and within their individual selves, while, at the same time, benefiting and sustaining the common welfare.

During FitzRoy's subsequent voyage to Tierra del Fuego with Darwin, the latter thought that one characteristic of Fuegian behaviour was "placid indifference,"[377] and he described a greeting between two Fuegians as being similar to "two horses in a field."[378] Darwin was unable to see that

---

373 C. McKew Parr, *Ferdinand Magellan, Circumnavigator*, (New York: Thomas Y. Crowell, 1964), 321.

374 E. F. Benson, *Sir Francis Drake*, (Edinburgh, UK: U Press, T. A. and Constable Ltd., 1927), 122–23.

375 Nick Hazlewood, *Savage: The Life and Times of Jemmy Button*, (London, UK: Hodder and Stoughton, 2000), 43-45.

376 Benyus describes an Amazonian native's lack of interest in Washington, D.C., where he went defend his homeland against oil drilling. "There is not much to learn in the city," he said, "it is time to walk in the forest again." (Janine Benyus, *Biomimicry: Innovation Inspired by Nature*, [New York: William Morrow and Company Inc., 1997], 9).

377 Janet Browne, *Charles Darwin: Voyaging. A Biography*, (New York: A. Knopf, 1995), 235.

378 Ibid, 251.

On Evolution

the pacific nature of the Fuegians may have been serenity, such a virtue in Twelve Step societies, rooted from childhood in the habitual practice of Higher Powered Mutual Aid. Raised in an authoritarian home in a rough English society, Darwin may not have recognized serenity, nor its basis or value.[379]

The Fuegian captive "Jemmy Button" was a favourite of the English, but Hazlewood says that there were "frequent references to his laziness" in journals kept with him in the Falkland Islands.[380] The European work ethic had perhaps not been instilled in Button, and he may have seen little need to work.[381] Furthermore, his "laziness" may have been related to his serenity, to his status as a captive, and/or being told what to do.[382] Citing observations of the Fuegians made in the early 1880s, possibly a subtle counter to Darwin's views, Kropotkin said in *Mutual Aid* that:

> [T]he Fuegians, whose reputation has been so bad, appear under a much better light since they begin to be better known. A few French missionaries who stay among them "know of no act of malevolence to complain of." In their clans . . . they share everything in common, and treat their old people very well. Peace prevails among these tribes.[383]

Like Darwin, Kropotkin did not know the history of the Fuegians, who had immigrated to inhospitable Tierra del Fuego after facing European persecution in their sacred lands to the north. According

---

379 My own initial reaction to the word serenity was anything but positive. To the "rowdyman" that I was, it was boring and unappealing. I later realized that serenity was precisely what I needed in my mad existence.

380 Nick Hazlewood, *Savage*, 205.

381 Asked the definition of work, an East Indies native once said it was "something the white man does."

382 Serenity is much valued in Twelve Step HPMA culture too, but being told what to do - given stern advice - is not. Rather, if a newcomer is relatively serene already, suggestion is much favoured.

383 Peter Kropotkin, *Mutual Aid*, 95.

even to Drake - in the early 1500s - the Portuguese had been savagely cruel to the Fuegians, driving them from their northern homelands.[384]

Fuegian serenity and mutual aid, then characteristic of First Nations cultures, may have been re-established over the years after Darwin's voyage in the early 1830s. This was evidently hidden to Kropotkin in his assessments of life in Tierra del Fuego. However, observing the serenity of other Indigenous peoples, Kropotkin also said:

> When first meeting with primitive races, the Europeans usually make a caricature of their life; but when an intelligent man has stayed among them for a longer time, he generally describes them as the "kindest" or the "gentlest" race on earth. These very same words have been applied to the Ostyaks, the Samoyedes, the Eskimos, the Dayaks, the Aleotes, the Papuas…by the highest authorities.[385]

In the late 1930s, Martin Gusinde studied and worked for six years in Tierra del Fuego, producing his voluminous *The Yamana: The Life and Thought of the Water Nomads of Cape Horn*. Jemmy Button was a Yamana or Fuegian.[386] Gusinde's views differ from those of Darwin. He says that the Yamana ". . . meet neither noisily nor excitedly [nor with] animated feelings of joy and friendship. We Europeans might interpret this…way of meeting one another as dullness if we were not certain of the sincere feeling behind it."[387]

Gusinde, like Darwin and Kropotkin, did not seem to know the history of the persecuted Fuegians, the loss of their northern sacred lands

---

384 John Hampden, ed., *Francis Drake, Privateer*, 1972, 137.
385 *Mutual Aid*, 91.
386 De Paolo notes that "Fuegian" was sometimes a misleading term and that under it were grouped three main tribes, the Yahgan ("actually Yamana"), the Alacaluf, and the Ona. (Charles De Paolo, *The Ethnography of Charles Darwin*, 2010), 43.
387 Martin Gusinde, *The Yamana: The Life and Thought of the Water Nomads of Cape Horn*, Translated from the German by Frieda Schutze, (New Haven, CT: Human Relations Area Files, 1961), 913.

and their resulting emigration to the south. Re-established over time in Tierra del Fuego, Fuegian emotional extremes were unnecessary in everyday life; thus, among them, there was no need to meet either "noisily" or "excitedly." Such behaviour would have been out of place in a society devoted to serenity and balance. Gusinde's work shows that Yamana life then was centred in respectful serenity; the result, to Gusinde?

> The entire social order of our Fuegians impresses one as kind and worthy of human beings [with] favorable mutual relations between parents and children, between the members of a kinship group, and among all the local groups.[388]

The recovery of the Fuegians, exiled from their northern sacred lands indicates a remarkable cultural resilience, a clear sign of how they operated so well in remembering and reconstructing Higher Powered Mutual Aid for themselves.

The neighbours of the Yamana (Fuegian), the Ona, were described by Lucas Bridges in his 1948 book, *Uttermost Part of the Earth*. Born and raised in Tierra del Fuego and Patagonia in the early 1900s, Bridges continued to live there as a missionary/naturalist for most of his life. He wrote a Yamana/English dictionary, and described First Nations life - not only in talking circles - that are reminiscent of Twelve Step meetings.

In the course of daily life among the Ona, said Bridges, "No one ever interrupted a speaker,"[389] and, even under duress, "they never interrupted one another or raised their voices unduly."[390] Though vast distances separated First Nations such as the Ona of South America and the Montagnais of the distant St. Lawrence River valley in North America, such cultures were similar in key respects such as these.

---

388 Ibid, 945.

389 Lucas Bridges, *Uttermost Part of the Earth*, (London, UK: Hodder and Stoughton, 1948), 200. Further, said Bridges on the same page, the Ona "did not shake or nod their heads [in conversation], for these signs of disagreement or consent were never employed..."

390 Ibid, 402.

Here, it is worth quoting Jesuit priest Lafitau regarding the conversational habits of the Wendat and Iroquois of North America, before extensive and intensive white influence on First Nations societies:

> In general, we may say that they are more patient than we in examining all the consequences and results of a matter. They listen more quietly, show more deference and courtesy towards people who express opinions opposed to theirs, not knowing what it is to cut a speaker off short, still less to dispute heatedly: they have more coolness, less passion . . . and bear themselves with more zeal for the public welfare.[391]

We can leave the last word to Paula Gunn Allen. She quotes Jesuit priest Lejeune as saying that the First Nations Montagnais people place "...all virtue in a certain gentleness..."[392]

Leadership in First Nations societies

Exposure to sophisticated forms of Higher Powered Mutual Aid in First Nations cultures drew different responses from various Europeans. The first of these, from people like Bridges and Gusinde, living and working in the southernmost parts of South America, are based upon extensive observation over time. Said Bridges:

> The Ona had no hereditary or elected chiefs, but men of outstanding ability almost always became the unacknowledged leaders of their groups. Yet one man might seem a leader to-day and another man

---

391 Georges Sioui, *For an Amerindian Autohistory: An Essay on the Foundations of a Social Ethic*, Translated from the French by Sheila Fischman, (Montreal: McGill-Queen's University Press, 1992), 48.
392 Paula Gunn Allen, *The Sacred Hoop*, 40.

to-morrow, according to whoever was eager to embark upon some enterprise.[393]

Similarly, one First Nations Elder said more recently:

> When the people gathered together [in the past], there was no one leader. When it happened that certain skills [were] needed, different people would take over. Everyone was equal and treated one another with respect, and all were humble towards each other.[394]

The authoritarian meaning of the word "sent," said Bridges, had to be ". . . used with discretion in relation to the Ona . . . nobody, were he magician or strongman, gave any orders, except, possibly, in actual attack."[395] Darwin noted, too, that New Zealand warriors fought without direction,[396] something that may have been a worldwide phenomenon with Indigenous peoples wedded to anarchic self-regulation and self-responsibility within the common welfare. Even when hunting, Black Elk said, it was "every man for himself."[397]

Among the Yamana, southern neighbours of the Ona, Gusinde made similar observations in a revealing account, saying:

> Ordinarily no one dare issue definite orders or make demands on other people, whether or not he is justified or obliged to do so . . . Even to this extent these [First Nations people] would not allow themselves to make demands or issue orders to a specific person. This was often observed. If I ordered a man to demand some service

---

393 Lucas Bridges, *Uttermost Part of the Earth*, 216.
394 Peter Kulchyski et al, eds., *In the Words of Elders*, 231.
395 Lucas Bridges, *Uttermost Part of the Earth*, 283, note 1.
396 Charles Darwin, *The Voyage of the Beagle*, (Hertfordshire, UK: Wordsworth Editions Limited, 1997, first published 1845), 398.
397 Black Elk, *Black Elk Speaks*, as told through John G. Neihardt, (Lincoln, NE: University of Nebraska Press, 1979), 56.

of another in my name, he listened to my words and departed, but he returned shortly thereafter without having spoken to the other man, and gave me to understand that I should personally give my instructions to the one concerned. In this way each person guards the other's freedom, and he also would not allow his own freedom to be curtailed.[398]

Citing some of Bridges's work, Gusinde offered an amusing anecdote, saying:

> The missionaries have had many experiences in this connection, for instance, when they provided...a supervisor for community work. "To be placed as overseer brought on at once a native jealousy and dislike. The office...was not popular [was] generally destitute of authority, and...fulfilled with laxity of purpose."...The natural inclination of these Indians is...that "they cannot bear to be dictated to"...as Bridges...tersely and definitely expresses himself.[399]

In a subsection titled "The Commonweal," Gusinde also describes non-authoritarian leadership in Yamana culture; it begins with the following:

> The Yamana are linked together in a definitely circumscribed area by a common language and a uniform way of life, but as a whole they are not under the administration of an authoritative person...nor are the various groups, large or small, led by a chief...[400]

A hundred years earlier, Darwin described the same peoples of Tierra del Fuego as savage and uncivilized, not knowing of the loss of sacred lands and their ensuing emigration into desperate straits at the southern tip of Americas. Kropotkin often used the word "savage" in *Mutual Aid*,

---

398 Martin Gusinde, *The Yamana*, 638.
399 Ibid, 638–39.
400 Ibid, 937.

On Evolution

though he sometimes placed it in quotation marks,[401] perhaps a subtle question posed to his readers. In its meaning as cruel, "savage" could be better used to describe the Europeans and their heartless drive for world domination.[402]

Darwin, professing to be an agnostic,[403] did not address spirituality in Indigenous societies extensively in his books, *The Origin of Species* and *The Descent of Man*. In *Mutual Aid*, Kropotkin was skeptical of the spirituality of Indigenous peoples.[404] Evidently, neither man recognized the central importance of guiding Higher Authority in these societies. Blinded by "scientific objectivity," by European social, economic, and political biases, and by superior attitudes, they also did not recognize the individual and community strength that Indigenous peoples drew from divesting themselves of authority while investing in humility.

Convinced of European superiority and that only hierarchical competition fuelled natural selection, Darwin concluded of the Fuegians that:

> [U]ntil some chief shall arise with power . . . it seems scarcely possible that the political state of the country can be improved . . . On the other hand, it is difficult to understand how a chief can arise

---

401 E.g., *Mutual Aid*, 83, 101, 103, 105.

402 Torrance says that the Portuguese were particularly savage when being ousted from Mozambique. (Tom Torrance, *Dar es Salaam, 1963*, [Renfrew, ON: General Store Publishing House, 2010], 113).

403 Two years before his death, Darwin wrote that he had "never been an atheist," and that "an agnostic would be the most correct description of my state of mind." May 7, 1879 letter to John Fordyce in the Darwin Correspondence Project, "Letter no. 12041." I accessed this on August 8, 2017.

404 Kropotkin portrays the mutual aid practiced in tribal and barbarian life in *Mutual Aid*, but does not make any firm spiritual analyses. He uses the word "god" in a few places (e.g., 106, 120, 133, 134), and tends to downplay or omit the importance of spirituality, e.g., in describing the sole authority of village assemblies, the "folkmotes" that he sometimes refers to (e.g., 142). Kropotkin was condescending towards Indigenous peoples in many places in *Mutual Aid*, when describing their ""thinkers" and *savants* - wizards, doctors, prophets, etc." (88, note 1), "a messenger from the sky" (106), "secret societies of witches, shamans, and priests, which we find among all savages" (114), and "wizards, priests" (121).

till there is property of some sort by which he might manifest and still increase his authority.[405]

Paula Gunn Allen describes the attempts of Jesuit priest Lejeune in the 1600s to convert the anarchic Montagnais to French ways of thinking and being. She says that Lejeune had a lot of work to do, to try to transform a people where children were not punished, where women were encouraged to be independent decision-makers,[406] where no one was held in authority, and who, in Lejeune's words, could not ". . . endure in the least those who seem desirous of assuming superiority over the others . . ."[407] Gunn Allen quotes Lejeune further to show his consternation when people rendered no homage to their Montagnais leaders, except if they wished to; said Lejeune:

> They have reproached me a hundred times because we fear our Captains, while they laugh at and make sport of theirs. All the authority of their chief is in his tongue's end, for he is powerful insofar as he is eloquent; and even if he kills himself talking and haranguing, he will not be obeyed unless he pleases [them].[408]

The Cherokee First Nation of the 1700s is described by Wright as "a loose federation of sixty-four autonomous towns united only by language, culture, and kinship. Chiefs led by example; bad ones simply lost their following."[409] Wright quotes a British officer as saying of the Cherokee:

---

405 Charles Darwin, *The Voyage of the Beagle*, (Hertfordshire, UK: Wordsworth Editions Limited, first published 1845, republished 1997), 219.
406 Gunn Allen noted the French-influenced shift from woman-centred society to male-centred society among the Montagnais First Nation. (Paula Gunn Allen, *The Sacred Hoop*, 40-41).
407 Ibid, 40.
408 Ibid
409 Ronald Wright, *Stolen Continents*, 94.

On Evolution

> There is no law nor subjection amongst them . . . the very lowest of them thinks himself as great and as high as any of the rest, every one of them must be courted for their friendship...So what is called great and leading men amongst them, are commonly old and middle-aged people, who know how to give a talk ...everyone is his own master.[410]

Sioui made similar observations in *Huron-Wendat: The Heritage of the Circle*. The essential quality of a chief, Sioui says, was the power to speak on behalf of the people he represented. The word "chief" - in the sense of someone who commands or directs - was unknown in First Nations societies. Chiefs were "only *representatives* chosen for their ability to act as a vehicle for...the voice of the people."[411] He also says that chiefs, "could never claim to be empowered to speak or act without consent on any matter whatsoever: they were spokesmen, nothing more."[412]

In his autobiography, *Where White Men Fear to Tread*, Sioux leader Russell Means said that, before the arrival of whites, First Nations did not have chiefs; they had leaders only, people ". . . chosen by consensus for their wisdom and courage." The concept of a "pyramidal hierarchy with a single person at the top," Means added, "was European." "When whites first demanded" to talk to a "chief," Means' forebears did not quite know what to say.[413] It should be noted here once more that Twelve Step HPMA groups have leaders who are only "trusted servants" who follow the guidance of the group.[414]

All this implies that leaders had to be good listeners, too, so that they could transmit the words of the people as oratory. And such was their love of individual freedom and making their views known, Sioui also observed,

---

410 Ibid

411 Georges Sioui, *Huron–Wendat*, 1999, 9.

412 Ibid, 128.

413 Russell Means, *Where White Men Fear to Tread: The Autobiography of Russell Means*, (New York: St. Martin's Griffin, 1995), 222.

414 See A.A.'s Tradition Two at the beginning of the book.

"Most Wendats possessed the gift of eloquence . . ."[415] He noted the same for the members of Iroquois society who learned "oratorical art . . . since for them . . . no one has arbitrary power over any other person."[416] Again, this is the same in Twelve Step fellowships.

Recognizing Iroquois cultural complexity, George-Kanentiio makes several comments that are consistent with the principle of non-human authority in First Nations societies. A memorable one, sounding strange to one raised in mainstream culture, is that "Iroquois children were taught to serve no master."[417] Sioui describes the similar teaching of Huron-Wendat children.[418] The opposite lessons are very often taught in much of the world today, to the detriment of children quashed by authoritarianism, to children's abilities, and to their self-esteem. Eventually they might become sick adults, too.

At about the same time that FitzRoy and Darwin were visiting Tierra del Fuego, the Frenchman Alcide d'Orbigny made his own exploration of the Americas. He observed that, in the Argentine Pampas (Patagonia), First Nations peoples could not understand the concept of a human master.[419] In Twelve Step culture too, as noted, no one person is above others: "leaders are but trusted servants," following the guidance of the group.

Lestringant describes a frustrated European perspective, saying that one basic quest for explorers was to find a "chief" to talk to, or someone who could ". . . express themselves in the name of authority . . . delegated

---

415 Georges Sioui, *For an Amerindian Autohistory*, 127–28.

416 Ibid, 47.

417 Doug George-Kanentiio, *Iroquois Culture and Commentary*, (Santa Fe, NM: Clear Light Publishers, 2000), 83. Fierce Lakota Crazy Horse, in the face of white pressure to capitulate, "knew that each one [of the Lakota] must do what seemed good for him." (Mari Sandoz, *Crazy Horse: Strange Man of the Oglalas*. [Lincoln, NE: University of Nebraska Press, 1961], 303).

418 Georges Sioui, *Huron-Wendat*, 154–55; Paula Gunn Allen, *The Sacred Hoop*, 216-17.

419 Alcide d'Orbigny, *Voyages dans des deux Amériques* (Paris: Furne, Jouvet and Co., 1857), 235.

On Evolution

by some chief or king."[420] Apparently in sympathy with explorers who could not find such chiefs, Lestringant says further, "One can . . . understand the confusion and even indignation . . . felt when facing the anarchy seemingly governing . . . peoples such as the Patagonians."[421]

In *The Journey of Crazy Horse: A Lakota History*, Marshall says that "...whites didn't understand, or simply ignored, the fact that there were different opinions among the Lakota."[422] He adds "...one or a few men couldn't speak for all the Lakota."[423]

Words from George-Kanentiio concern leadership in Iroquois society: ". . . the most important quality any Iroquois leader can have," he said, "is a lack of personal ambition with regard to power and prestige. Iroquois customs dictate keeping power away from any individual who actively seeks it."[424]

Again, this seems strange to the mainstream mind. We elect those who want to be in power, seldom fully understanding their motives. Those who seek power may wish to be in power, in control, because, to a great extent, they think that they know what is best for others. They can be good orators but they are not necessarily good listeners, nor are they necessarily good vehicles for the voices of the people who elect them.

Though weakened, First Nations HPMA principles remain today. The First Nations characteristic of "...strict adherence to the principle of self-determination..." has been considered a major problem for social work practitioners attempting interventions in First Nations societies.

---

420 Frank Lestringant, "The Myth of the Indian Monarchy: An Aspect of the Controversy between Thevet and Lery," in Christian Feest, editor, *Indians and Europe: An Interdisciplinary Collection of Essays* (Lincoln, NE: University of Nebraska Press, 1989), 37.
421 Ibid
422 Joseph Marshall III, *The Journey of Crazy Horse: A Lakota History*, (New York: Viking, 2004), 206.
423 Ibid, 208.
424 Doug George-Kanentiio, *Iroquois Culture and Commentary*, (Santa Fe, NM: Clear Light Publishers, 2000), 129.

James G. Duncan

[425] During a sweat lodge ceremony, I suggested that a First Nations man and myself should adopt separate roles as firekeepers, heating the rocks where they could be moved, red hot, into the lodge upon the request of its leader. The man listened to me and firmly offered another view, that I should adopt the role I had suggested for him!

Fear and First Nations HPMA

The sun pours down on . . . the lovely land that
man cannot enjoy. He knows only the fear in his heart.

ALAN PATON

Related to and interwoven with serenity and trust is their counterpart, fear. First Nations Higher Powered Mutual Aid societies may have been adept at transforming fear into serenity and trust before the Europeans made contact with them. However, Europeans invading the Americas had cultures *founded* in fear and mistrust, authoritarian hierarchy, imperialism, and war, having long ago forgotten or abandoned HPMA. Cultures built for war are superior at producing as well as conducting war, not peace.

As far back as Magellan, the Europeans both exemplified and used fear, and serenity has taken a plunge throughout the Americas for over 500 years. The Europeans had begun trying to break links in the ecological circle, and were thus essentially ill. The ecological circle can never be broken, actually, but we continue to try, bringing us ever closer to oblivion. Whites attempted to step outside the circle, to live in anti-ecological illusion, to protect themselves from the effects of their unremitting "war of all

---

[425] J.G. Good Tracks, "Native American Non-interference," *Social Work*, November 1973, 30.

against all."⁴²⁶ They were fearful and unable to remember the ecological underpinnings of their lives, namely that the locus of creative power is within the free, self-regulating, self-responsible individual in egalitarian, ecological community, not in authoritarian hierarchy.

MAGELLAN THEN DRAKE IN TIERRA DEL FUEGO

Joyner records Magellan's use of brutality and executions against his men, and the establishment of fear and anger in the First Nations Patagonians. Wanting to bring some of the huge men back to Spain as "curiosities," Magellan ordered his crew to capture two of them. Using deception and trickery, they loaded the trusting men high with gifts, then, disarmed, they were put in leg irons and taken on board Magellan's ship.⁴²⁷

Drake's own use of barbarity and fear indicated that he and Magellan were cut from similar cloth. The English captain ordered the execution of one of his seamen, and, according to Kelsey, he purposely wished to have his men fear him.⁴²⁸ However, Drake found in the end that, not only were his men afraid of him, but he was afraid of his men. Fear ruled his entire ship, and it is not surprising that the instilling of fear among First Nations peoples was one result.

Accounts of Magellan's voyage were known by Drake and Fletcher, who thought that, after the former's contact with them, the Patagonians were fearful, with white pressure from the north, and now from the south. A remarkable passage shows the importance of liberty to the Patagonians; Drake (or his chaplain, Fletcher) said:

---

426 Thomas Hobbes in *Leviathan*, 1651. "The natural state of mankind [before a central government is formed] is," said Hobbes, a "warre of every man against every man." Darwin had such European ammunition for his books on evolution, *The Origin of Species* and *The Descent of Man*.

427 Tim Joyner, *Magellan*, (Camden, ME: International Marine, 1992), 148-49.

428 H. Kelsey, *Sir Francis Drake: The Queen's Pirate*, (New Haven, CT: Yale University Press, 1998), 113.

> [D]uring Magellan's passage], they lived as a free people...but now are in most miserable bondage and slavery, both in body, goods, wife and children, and life itself to the Portugals, whose most hard and cruel dealings against them forceth them to fly into the more unfruitful parts of their own land, rather there to starve, or...live miserably with liberty...[429]

This is another instance of the flight of First Nations people, the Patagonians this time, to "unfruitful" land, to escape the onslaught of whites in the north *and* south. An observant Drake (or his chaplain) knew of this exile from ancestral lands, but Darwin, Kropotkin, Bridges, and Gusinde apparently did not know of it among the Fuegians.

Drake's account resumes with a good spin. After initial fear, he says, the Patagonians later "rejoiced greatly in our coming, and in our friendship, in that we had done them no harm."[430] At another harbour on the southeastern coast of Patagonia, Drake's record shows that "the people of the country did show themselves to us," and their contact was evidently friendly. Drake (or Fletcher) made a beautiful description of these people who had not yet had bad experiences with Europeans:

> They are clean, comely, and strong bodies; they are swift of foot, and seem very active . . . It is wonderful to hear, being never known to Christians before this time, how familiar they become in short space with us; thinking themselves to be joined with such a people, as they ought rather to serve than offer any wrong or injury unto . . . neither seemed their love less toward us.[431]

But some Patagonians seem to have remained wary. Further south along the coast, two men "of huge stature and strength" showed friendliness

---

429 John Hampden, ed., *Francis Drake, Privateer*, (London, UK: Eyre Methuen, 1972), 137.
430 Ibid, 140–41.
431 Ibid, 143–44.

to the English, but when an English bow appeared, perhaps in fear, the picture changed abruptly and a terrible battle took place, with loss of life on both sides.[432] Drake reflected perceptively:

> [T]his is certain, that the...cruelties there used have made them more monstrous in mind and manners than they are in body, and more inhospitable to deal with any strangers that shall come hereafter. For the loss of their friends [through Magellan's treachery] (the remembrance whereof is assigned and conveyed over from one generation to another among their posterity) breedeth an old grudge, which will not easily be forgotten . . .[433]

What may not have been forgotten also was the tendency of the Europeans to change tactics to gain advantage. Memories of Magellan's trickery no doubt made the Patagonians more fearful and prepared, as Drake (or Fletcher) surmised. However, the Englishmen may also have failed the test of fear with the huge Patagonian men by allowing the appearance of the powerful English bow.

Darwin aboard the *Beagle*

As with captains Magellan and Drake, 300 years later there was fear of captain FitzRoy's punishments among his sailors.[434] Charles Darwin was seasick from the trip's beginning; Browne describes what might have been the combined effect on him:

> Swirling around with the seasickness in Darwin's head were shrieks from sailors being flogged...Four sailors were beaten with the cat-o'-nine-tails: one for drunkenness...one for disobeying orders... one for drunkenness and insolence... and one for neglect of duty... Darwin barely possessed the energy to know what he thought about

---

432 Ibid, 146–47.
433 Ibid, 147.
434 Adrien Desmond et al, *Darwin*, 1991, 115, 120.

this except that he had somehow stepped into an Old Testament version of hell: the screams mingled with his own nausea in a sickening whirlpool of revulsion.[435]

Fear of the sea was a factor for all in their tiny ships, sailing off into unknown waters.[436] Raised in the illness of hierarchical authoritarianism, the sick Europeans saw nothing wrong, again, with their continuing efforts to sever links in the chain of the ecological circle. That disease among Old World peoples was associated with overpopulation, giving them an excessive ecological footprint. Poverty, homelessness, physical and mental illnesses, and perhaps alcoholism, had become common amidst affluence.[437] They lived what they were and transferred their illnesses to others.

As noted, Darwin and many others, including his rival A.R. Wallace, had been strongly influenced by clergyman T.R. Malthus. So influential was Malthus that his:

> ...followers in the 1830s circumvented death by deportation and shuffled the poor out of the country...Huge numbers were leaving as the Depression deepened, 400,000 annually - to America, Australia, the Cape...Indeed, '[all the] *unpeopled* regions of the earth may now be said to be British ground.'[438]

In his comparison of the English and Fuegian languages, Hazlewood notes that the Fuegian *"yammerschooner"* had long been interpreted as "give me." However, this proved to be incorrect for hundreds of years, and its actual meaning speaks volumes about Indigenous peoples' perceptions of the Europeans. Hazlewood says that for centuries the Natives "had been

---

435 Janet Browne, *Charles Darwin: Voyaging*, 167–68.
436 Iain McCalman, *Darwin's Armada* (London, UK: Simon & Shuster, 2009), 49.
437 Indigenous visitors to Europe were reportedly shocked at the poverty amidst European affluence. (See Paula Gunn Allen, *The Sacred Hoop*, 219).
438 Adrien Desmond et al, *Darwin*, 1991, 266. Italics mine.

On Evolution

abused by passing ships and had feared the presence of strangers and the potential for violence that they brought. *Yammerschooner* did not mean "give me" but "be kind to me," or "be kind to us."[439]

The Fuegians (Yamana) cannot teach us now about the meaning of kindness; most tragically, they are no more. Societies underlain by benign anarchic groups can produce serenity - peace - not an ability to resist authoritarian hierarchies founded in and employing fear, and built for effective war. Despite Darwin's remark that the Yamana language "scarcely deserves to be called articulate,"[440] it may have been more sophisticated than that of English. Bridges's Yamana-English dictionary defined sixty-one words for social relations in Yamana, well over twice the twenty-five for English.[441]

---

439 Nick Hazlewood, *Savage*, 325.
440 Charles Darwin, *The Voyage of the Beagle*, 196.
441 Nick Hazlewood, *Savage*, 324.

CHAPTER 5

First Nations and Twelve Step HPMA

The downfall of humanity began with the loss of its humility.

ANONYMOUS

THE AMAZING SET of principles and tools that we have at our disposal to learn from ourselves, and in turn from each other, is Higher Powered Mutual Aid. First Nations have been practicing it for a long time, and pioneering Americans from the northeastern United States unearthed it in their struggle against alcoholism during the 1930s. A description of parallels between First Nations circles and Twelve Step groups includes the physical and spiritual settings for the circles and groups, guidelines for speaking (sharing), characteristics of circles and groups, and some of the key effects of "working the program" in these societies that are similar in such important respects.

Settings

Humans evolved together in small groups, the natural, basic structures of HPMA societies. They were the most important place where HPMA principles were learned and put into practice. A clear, healing communication can take place in groups that offer equality, safety, and serenity for all.

Sioui uses the word circle to describe non-hierarchical First Nations societies, particularly in *Huron-Wendat: The Heritage of the Circle* (1999). Many aspects of First Nations HPMA societies are organized around the spiritual and ecological principle that everything is interdependent,

On Evolution

in a line that forms a circle or circles, that is, a sphere. Gunn Allen says tribal people consider that the human and non-human worlds are linked together within "one vast, living sphere,"[442] like the Earth itself. Indigenous peoples have always been spiritual ecologists, but the diseased, hierarchically-organized Europeans could only describe them pejoratively as pagan, primitive, savage, and/or uncivilized.

That the circle is a powerful entity was not lost on the Twelve Step pioneers. In 1957, Bill W. observed that the circle enclosing an equilateral triangle had been regarded by the ancients as a means of defence against evil spirits.[443] A.A. adopted it as a symbol for the three Legacies of Recovery, Unity, and Service,[444] each on one side of the triangle, itself protected within a circle. The circle enclosing a triangle was registered as an A.A. trademark from 1955 to 1993, and its use is still prominent in A.A., for instance, in pamphlets such as "Circles of Love and Service." Al-Anon has enclosed the circle inside the triangle to derive its own symbol.

Bill W. did not record parallels between Twelve Step groups and First Nations circles, and the idea of the ecological interdependency of all things, though valid, went far beyond the singleness of purpose that A.A. needed to adopt to deal with sometimes life-threatening alcoholism. Yet the Twelve Step pioneers left wide open the door to a complete, ecological approach to Higher Powered Mutual Aid, particularly in the Twelfth Step. We address this in more detail in Chapter Six.

The door or gate is also a powerful symbol in Twelve Step HPMA, as it is among First Nations. First Nations medicine wheels have doors to the four directions, and circles and sweat lodges have gates that open to the east, whence comes the sun. In *Alcoholics Anonymous Comes of Age*, Bill W. referred to the widening of the A.A. gates, e.g., through valuable contributions from atheists and agnostics, "...so that all who suffer might pass through, regardless of their belief or *lack of belief.*"[445]

---

442 Paula Gunn Allen, *The Sacred Hoop*, 22.
443 *Alcoholics Anonymous Comes of Age*, 139.
444 Information about the Legacies can be found in A.A. and Al-Anon literature.
445 *Alcoholics Anonymous Comes of Age*, 167.

James G. Duncan

In Twelve Step HPMA, those having difficulty with the concept of a Power greater than themselves - egos can be very large - are in good company.[446] Bill W. had such difficulty, and a whole host of atheists and agnostics made their way in the door, and still do.[447] Says Nowinski, Bill W. was a "lifelong"[448] and "avowed"[449] agnostic himself.

Former atheist Nancy Abrams came into the Twelve Step tent when she realized that she needed a Higher Power.[450] She says that "having the explicit nonreligious moral principles of the twelve steps has made my life much easier."[451] She wonders if we can "...learn from the twelve steps how valuable it might be to have a set of nonreligious moral principles that could undergird decisions with planetary consequences."[452] Yes we can.

A.A.'s *Came to Believe* . . . (1973), Al-Anon's *As We Understood* . . . (1985), and Al-Anon's *Having Had a Spiritual Awakening* . . . (1998), show how broad Twelve Step spiritual doors are. Higher Powers defined as the door to a meeting room, and the unofficial slogans, "take what you like and leave the rest" or "fake it 'til you make it," often help people get in the door and keep them coming back to get, and give, help. Many are desperate and creative enough as newcomers that it is not terribly difficult to enter into this new world with a door-like Higher Power until they develop another, if they so wish. There might be *acceptance* of the idea of something greater than oneself. In 1934, Bill W. learned

---

446 In *Sane: Mental Health, Addiction, and the Twelve Steps*, 2010, Marya Hornbacher describes the difference between healthy self-esteem, and arrogance or egotism (27). "Getting down to right size," she says, can be achieved by being humble and mature in relation to others and to our Higher Powers (41).

447 Ibid, 167.

448 Joseph Nowinski, *If You Work It, It Works! The Science Behind 12 Step Recovery*, 2015, 20.

449 Ibid, 138.

450 Nancy Ellen Abrams, *A God That Could Be Real: Spirituality, Science, and the Future of Our Planet*, (Boston: Beacon Press, 2015), xix.

451 Ibid, 132.

452 Ibid

On Evolution

from Ebby T. that all he need do was to believe in an entity greater than himself.[453]

Lest the idea of a Higher Power as a door be seen as disrespectful or frivolous, I suggest that it has a powerful practical and spiritual application. Individual comments in Twelve Step speaker and discussion meetings often draw sharp distinctions between what goes on *inside* the doors of HPMA meetings and what goes on *outside*, in the really crazy world that we live in. Outside the doors of meeting rooms there is hierarchical human authority; inside the doors, only the authority of benevolent Higher Powers is recognized. Thus, the door represents access to the freeing of the heart, mind, and spirit. The anarchic absence of rules for individual recovery within the group or circle, behind the doors and under the authority of Higher Powers, is an essential component of what defines and makes effective the benign anarchy of Twelve Step groups and First Nations circles.

Before the arrival of Europeans in the Americas, replete with authoritarian hierarchy, First Nations societies employed HPMA, integrally, as part of their daily lives. Forced to go underground for many years,[454] these traditions are now being brought back. As in Twelve Step societies, First Nations now employ more meetings that can be open to all, whether of First Nations heritage or not. Thus, with much promise First Nations gates are widening, too. An important turning point in my life was my acceptance into a First Nations Men's Healing circle. There I first found parallels between First Nations and Twelve Step Higher Powered Mutual Aid.

A comparison of discussion guidelines in First Nations circles and Twelve Step groups, and some characteristics, reveals remarkable similarities between them. Recorded from almost three decades of personal experience and study are the following data.

---

453 *Alcoholics Anonymous* (the Big Book), 12.
454 See, for example, Peter Nabokov, *Native American Testimony*, 213–31.

James G. Duncan

A Comparison of First Nations circles and Twelve Step groups

	First Nations[455]	A.A	Al-Anon
Common Welfare 1st	Yes	Yes	Yes
Spirituality	Yes	Yes	Yes
Mutual Aid	Yes	Yes	Yes
Individuality	Yes	Yes	Yes
Focus on self	Yes	Yes	Yes
Interrupt speaker	No	No	No
Confidentiality	Yes	Yes	Yes
Suggestion favoured	Yes	Yes	Yes
Obliged to speak	No	No	No
Dominance	No	No	No
Challenge others	No	No	No
Flexibility	Yes	Yes	Yes
Rotating chair	Elders may	Yes	Yes
Children included	Sometimes	No	No

The guidelines and characteristics of First Nations circles and Twelve Step groups depicted in this table contribute to a sense of safety, so that all participants may listen and speak without the barrier of fear.

This leads into consideration of what I call the "sleepers of safety" that make HPMA circles and groups work. Often heard at Twelve Step meetings is a phrase like, "I don't know how it works, but it works." What I am talking about is different from the fifth chapter of the Big Book, "How It Works," or the Al-Anon book, *How Al-Anon Works*. It is a brief, specific account of how First Nations circles and Twelve Step groups function to promote morality and instill feelings of individual freedom, safety, and trust, all within the common welfare.

---

455 First Nations circles are different from each other in certain respects, less important for our purposes.

On Evolution

The better part of my HPMA experience has been in Twelve Step groups, and the discussion will reflect this. Nevertheless, I believe that this is applicable to circles in First Nations societies as well. I emphasize that I will describe this only briefly, partly because it is only through subjective experience - especially participating in circle and group meetings - that one can truly learn and use the principles of Higher Powered Mutual Aid. Attending A.A. meetings is recommended for the clients of analyst Marion Woodman,[456] and for Anne Wilson Schaef's trainees.[457] Barbara Wood counsels adult children of alcoholics to attend Al-Anon meetings.[458]

The following section describes some critical Al-Anon discussion guidelines, which, when thoroughly followed, have beneficial effects operating mainly below the level of immediate, conscious awareness. These are almost identical to guidelines in First Nations circles and A.A. groups, the main difference being that guidelines in the latter societies are most often unspoken, learned by the listening watchful through participation.

The Sleepers of Safety

Some group guidelines and characteristics, together with the process they describe, are portrayed here as necessary for the individual good and the common welfare. HPMA meetings provide a safe place, a haven, where people can let down their guard or take off their masks, trust, and become who they are and who they can be. Contrasting the atmosphere *outside* Al-Anon meetings with the beneficial atmosphere established by the guidelines and characteristics *inside* meetings, one quiet newcomer,

---

456 Marion Woodman, *Addiction to Perfection* (Toronto: Inner City Books, 1982), 28.

457 Anne Wilson Schaef, *When Society Becomes an Addict* (San Francisco, CA: Harper and Row, 1987), 145; Schaef, *Beyond Therapy, Beyond Science: A New Model for Healing the Whole Person* (San Francisco, CA: Harper and Row, 1992), 272.

458 Barbara Wood, Children of Alcoholism (New York: New York University Press, 1987), 5.

surprised, said, "It blows me away!" Due to its importance, we can examine this "sleeper" process in more detail, with reference to key guidelines.

ALLOW ALL TO SPEAK WITHOUT INTERRUPTION

Not being interrupted lets us know that we are being listened to, and we feel more whole, more worthy. Others respectfully give us time to say what is on our minds and in our hearts. First Nations circles often have a sacred feather, stick, or stone passed around to denote the speaker. But when we are deep in thought, or searching for words to express them, an interruption can sever the thought like a knife. Such interruptions occur rarely in First Nations circles and Twelve Step groups, but when they do, results are predictable: speakers can lose their train of thought. In fear, the thought finds refuge in silence.

As a child enmeshed in the disease of alcoholism, and in strict authoritarianism at home and school, I was often not allowed to speak, or my words were curtailed, interrupted, or discounted. I thus learned unconscious and destructive low self-esteem. The listening and speaking I did then was valuable partly for reasons of personal safety. I did a great deal of listening and speaking in fear, fear that I lived with later for much of my life, some of it conscious but mostly hidden in the unconscious. My experience as a child is little different from that of others in HPMA circles and groups, even that of others raised in mainstream hierarchical families and communities.

However, in First Nations circles and Twelve Step groups, I learned that what I said *was* important to me and to others. Long participation in First Nations and Twelve Step HPMA helped me to make conscious the unconscious fears and associated harsh feelings accumulated from childhood, and I stood at the threshold of greater liberty.

DO NOT COMMENT ON WHAT OTHERS SAY

Not having what I say in meetings qualified, debated, contradicted, or criticized, allows me to define my very own place, purpose, and truth in the world. Furthermore, this showed me that group members would respect others' words. Gradually, with the aid of the program and a gifted psychologist, I was able to find a core of myself around which I could build and make a life. In practice, that core - the real me - became recognizable only after boundaries began to form, and personality reconstruction became a bit of this and that, to get by, with faith, hope, and prayer.[459]

One day, I arose with the realization that change, a new consciousness, had set in, had taken place in HPMA, and I knew why. It was the benign anarchy devoid of rules that transformed, as in a dream, with guidelines for respect for the self and for the other. Bill W. wrote of the new consciousness in 1953.[460] Such is the progress of the Twelve Step HPMA movement that Room optimistically writes, as earlier noted, of a generalized Twelve Step consciousness in the world today.[461]

KEEP THE FOCUS ON YOURSELF, NOT ON OTHERS

This is a difficult one, especially for newcomers focused on the harm done by alcoholism and other illnesses. It is almost impossible to move quickly from the past to the present and on to brighter tomorrows. Suggested guidance is patient and empathic, from First Nations Elders and Twelve Step sponsors. As newcomers learn to focus on themselves, they begin to listen and talk through their pain, and learn to use the tools of their HPMA programs to arrive at a place of sobriety, sanity, serenity, improved morality,[462] self-regulation, forgiveness, respect, and love, in all their

---

459 I've always had some faith in myself, a certain self-love probably given me as a child. This has allowed me to bear burdens and to move on, sometimes blindly, to a semblance of wholeness.

460 *Twelve Steps and Twelve Traditions*, 107.

461 Robin Room, "Healing Ourselves and Our Planet: The Emergence and Nature of a Generalized Twelve-Step Consciousness," *Contemporary Drug Problems* 19, no. 4 (1992): 717–40.

462 Marya Hornbacher discusses morality extensively in her book, *Sane*, 34-115.

affairs. Discussion in First Nations circles and Twelve Step groups helps keep the focus on the self.

Entreaties by Twelve Step oldtimers not to talk about the past may serve to turn newcomers away. Encouraging the alcoholic home "don't talk" rule can keep us mired in the family disease of alcoholism.[463] Yet continually dwelling on the past can also keep us mired in it, curtailing progress in recovery. Al-Anon might benefit from a more consistent use of "living with alcoholism" rather than "living with an alcoholic." This could help to keep the focus on ourselves and the family disease.

KEEP WHAT YOU HEAR CONFIDENTIAL

People who are baring their hearts and souls, their private lives, in the presence of others, need to know that "what is said here, stays here." Gossip and criticism are not okay, and taking others' inventories is a pastime for those not working their own programs in productive ways. In this manner, people can feel a sense of safety and trust that they may not have known for some time, if ever. If confidential matters are aired outside the circle or group, or if gossip about or criticism of others takes place, much harm can result.

With the group chairperson or circle Elder and others employing self-respect, respect for others, and respect for the HPMA guidelines and process, one eventually feels free to listen and speak without fear, able to let one's guard down, to trust, to realize that others would not criticize what one has to say. Further, listening to others is not impaired by the formulation of counterargument: these are not debating societies. This is crucial to recovery and healing in HPMA societies, and to the individual liberty within the common welfare that they provide.

Leading meetings is a responsibility whose significance is sometimes underrated in Twelve Step HPMA, partly because it is carried out more casually or informally. Elders providing leadership in First Nations circles

---

463 The Al-Anon pamphlet, *Alcoholism, The Family Disease* (1996), briefly describes how alcoholism affects the families and friends of drinkers.

are usually well versed in HPMA principles, especially spirituality. Adequate leadership will help to keep members coming back, both newcomers and oldtimers. This is a critical element for individual recovery within group and circle atmospheres rendered safe by adherence to the discussion guidelines.

If we are to truly learn better habits of living, to carry the message with us "in all our affairs,"[464] rooted in sobriety, sanity, serenity, respect, and trust, it is the observance of the principles in practice that will help guide us along. Eventually, the benefits of "living the program" begin to shine, self-esteem and self-regulation shoot up, unhealthy dependency and immorality[465] take a plunge, and recovery starts to set in with a will.

Spiritual Parallels

Encouraging words from a First Nations perspective support the idea that HPMA is quite similar in First Nations circles and Twelve Step groups. Brian Maracle's *Crazywater: Native Voices on Addiction and Recovery*, quotes a First Nations woman on the return to freedom that she found in A.A., and A.A.'s compatibility with her culture. Raised in First Nations spiritual traditions - similar to those of A.A. - the woman said that she'd lost them all through alcohol abuse.[466] She explained, "I was a free person...I was born with that freedom. And in AA I got that back...My spirit died [through alcohol abuse]. It came back to life through the AA program."[467]

---

464 The phrase "in all our affairs," ending the Twelfth Step, is crucial, as explained in Chapter Six.

465 A.A.'s Steps Four through Ten deal with the adoption of a new and improved morality.

466 Brian Maracle, *Crazywater: Native Voices on Addiction and Recovery* (Toronto: Viking, 1993), 261.

467 Ibid, 262.

James G. Duncan

Coyhis describes A.A.'s profound usefulness in dealing with alcoholism among First Nations peoples.[468]

A.A. published a pamphlet called "AA for the Native North American" (1989), with brief sharings from First Nations people. Here, "Baba C." said how happy she was when she found out that she did not have to give up her ancestral beliefs to join A.A. She, like all others, could have the Higher Power of her own understanding. Participating in the A.A. program in this way actually strengthened her belief system.[469]

Makela also notes the cultural adaptability of A.A., citing, among others, the example of the Coast Salish First Nation of British Columbia, Canada.[470] *AA around the World* describes Twelve Step HPMA's compatibility with other First Nations societies. For instance, because of their own strong spiritual beliefs, the Australian Kooris[471] and the New Zealand Maoris[472] have found A.A. to be very valuable in their healing work.

In its own literature, Al-Anon recognizes parallels with First Nations societies. An Al-Anon member's sharing, called "A Revitalizing Power" in *Having Had a Spiritual Awakening...*, is similar to that of Baba C. in A.A., above, but is more descriptive. Comparing Al-Anon and First Nations teachings, the writer said:

> I have found many similarities in the teaching of Al-Anon and the teaching of respect for others that is the heartbeat of my heritage...

---

468 Don Coyhis, *The Red Road to Wellbriety - in the Native American Way*, (Aurora, CO: Coyhis Publishing, Inc., 2002); with William White, *Alcohol Problems in Native America: The Untold Story of Resistance and Recovery --"The Truth about the Lie"*, (Aurora, CO: Coyhis Publishing, Inc., 2006); Coyhis, *The Wellbriety Movement Comes of Age: The Fulfillment of Prophecy*, (Aurora, CO: Coyhis Publishing, Inc., 2011).

469 Alcoholics Anonymous, "AA for the Native North American," pamphlet, (New York: A.A. World Services, Inc., 1989), 6.

470 Klaus Makela et al, *Alcoholics Anonymous as a Mutual-Help Movement*, 27.

471 *AA around the World: Adventures in Recovery* (New York: The AA Grapevine, 2000), 50-52.

472 Ibid, 101-103.

On Evolution

My ancestors usually hold their ceremonies in circles...In my Al-Anon home group, we have our tables and chairs set in a circle, and I feel a spiritual connection as I look around into others' eyes... When one of us speaks of his or her Higher Power...I feel a surge of energy and a revitalizing power fill the room. I feel it again as we hold hands in our closing circles. I have read about and talked to many Native Americans who recount this same phenomenon in their celebrations and clan gatherings. They say they get their individual strength from the coming together of their clan members, and it sustains them until they meet again...I feel the same about my Al-Anon meetings! I feel a wonderfully ancient yet ever-youthful universal force is at work in our world and my heart leaps for joy that I am a part of it.[473]

Children

One evening at a First Nations circle a boy of eight years appeared. He was quiet and seemed beyond his years in serenity and wisdom. I saw him in circle again when he was 11. One First Nations Elder said of participation, "Everyone has their time to speak, even children were asked for their opinions"[474] and "Every child was encouraged to develop self-expression."[475] The boy's participation in the group told me that a family approach in Twelve Step groups might be helpful again if it was adequately supported.

It is not often noted that early in Twelve Step history, children were included in group meetings, for instance in the home of A.A.'s cofounder, Dr. Bob, and his wife, Anne S.[476] Twelve Step meetings sometimes do have

---

473 Al-Anon Family Groups, *Having Had a Spiritual Awakening . . .*, (Virginia Beach, VA: AFG Headquarters Inc., 1998), 126.
474 Peter Kulchyski et al, *In the Words of Elders*, 73.
475 Ibid, 80.
476 See Bob Smith et al, *Children of the Healer: The Story of Dr. Bob's Kids*, (Park Ridge, IL: Parkside Publishing Corporation, 1992).

children, when for example child care is not available, but not generally for contributions that they might make to the group discussion. Alateen, founded in 1957 as a part of the Al-Anon Family Groups, is not yet as developed as it might be, nor was Alatot a success.

The presence and active participation of children at some Twelve Step meetings could be refreshing and illuminating, allowing us to see the disease through younger eyes. We should be comfortable with what comes out of the mouths of babes. The young may not yet have learned fear and artifice, and what they say is often the unvarnished truth. The truth is less painful in supportive HPMA groups. Stephanie Brown et al provide details of their work with A.A. and Al-Anon members, showing the need to include children and suggesting some means for this to happen.[477]

More attention might be paid to rites of passage from childhood to adult. Here, I am referring specifically to the passage into adult life which has a fuller range of choices, where children/young adults learn new responsibilities to and for themselves, and to others in the world around. This is a huge, underrated responsibility, to the degree that the community needs to be involved as well: "it takes a village to raise a child"[478] is all too true. Joseph Marshall III says that in First Nations Lakota society every adult was indulgent towards children and gave them all the attention they wanted.[479]

In authoritarian families, not just in alcoholic homes, children's range of choices is limited, sometimes extremely so. Scott Peck described his childhood, saying that he was "not free to be me."[480] As adults, children of alcoholics are faced with all kinds of difficult choices for which they may not be well prepared. Twelve Step societies have yet to get a good grip on this subject, and it is worthy of detailed treatment elsewhere.

---

[477] Stephanie Brown et al, *The Alcoholic Family in Recovery*, (New York: The Guilford Press, 1999). Some parents are using the Twelve Step model for children using drugs (CBC Radio 1, October 13, 2017).

[478] A Nigerian proverb.

[479] Joseph Marshall III, *The Journey of Crazy Horse*, 47-48.

[480] M. Scott Peck, *The Different Drum*, 1987, 27.

On Evolution

For children in authoritarian, hierarchical homes and societies, deprived of adequate guidance and robbed of self-responsibility, an awkward trial-and-error approach to decision-making may begin at puberty, e.g., with drinking and sexuality. Unceremonious passage into adulthood can leave the child hiding within the body of an adult, never having been recognized, and never recognizing a purpose for existence. If they then become parents and try to teach their own children about self-regulation and self-responsibility, they may be using a recipe for disaster.

However, stepping out into a safe circle of family and community, guided by experienced people, and by Higher Powers, children can prepare for life as individuals with healthy purpose(s), for the self in relation to the other. Children can be given more choices as they are able to assume them, following the example of adults in non-authoritarian milieus. A thorough investigation of the possibilities is beyond the scope here, but there are already some good examples of how to go about this.

Twelve Step societies recommend that newcomers get sponsors to guide them through their programs. First Nations people find guidance from Elders. The same might be done with child newcomers. A.A. pays attention to the passage from drinking to sobriety through the "chip system," making announcements of sobriety dates, and so forth. Some Al-Anon groups celebrate the dates of arrival in the program, "birthdays." Newcomers' meetings in A.A. and Al-Anon are an initiation to Twelve Step HPMA.

Perhaps some form of First Nations "vision quest" is needed in mainstream society to assist youth in determining purpose earlier in life. This is far, far preferable to turning young adults loose with poor or insufficient guidance from their Elders and sponsors, without the inner compass that HPMA can provide, into a confusing world where an HPMA program is necessary for healing and recovery down the road. Many cultures have important ceremonies and celebrations at puberty, including those of First Nations,[481] and it might be beneficial for wider society to adopt such approaches.

Ecology

A key difference between First Nations and Twelve Step HPMA societies concerns connections with the rest of Creation. First Nations societies are very much rooted in all of Creation, and, though affected by outside culture, their circles today are still mainly rooted in ecological and spiritual traditions. Kulchyski et al say that First Nations Elders "teach a world-view based on the knowledge that all things in life are related in a sacred manner...Mother Earth is...held to be sacred, a gift from the Creator."[482] One Elder felt that "...Mother and Earth are one word - Motherearth."[483]

Twelve Step HPMA groups stay very close to the purposes for which they were designed: recovery from often grave illnesses. Other issues, such as ecology, the causes of substance abuse, politics, and religion, are outside the purposes of Twelve Step HPMA. If such issues were to become the focus of discussion, attention would be diverted from a life's work of healing and recovery. Yet, the Twelfth Step's suggestion to practice the principles "in all our affairs" leaves wide open the door where we may enter, into a place of greater respect for ourselves, for the common welfare, and for all of Creation. It should be understood that this includes ecological interrelations with our "Motherearth."

With our reconnection to all that exists, we lay down strong foundations for a thoroughgoing HPMA in our lives. While Twelve Step groups must cleave to their single purpose, individuals or groups can take a comprehensive approach to the latter part of the Twelfth Step,

---

481 See for example Peter Kulchyski et al, *In the Words of Elders*, 75, 154, 371, 407, 429-30.
482 Ibid, xvi.
483 Ibid, 91.

practicing "these principles in all our affairs."⁴⁸⁴ Here, all our affairs includes relations with all of God's Creation. Al-Anon recognizes this in a daily reader, referring to the great potential of the program in relations with all of life.⁴⁸⁵

The thought and action of natural, holistic First Nations HPMA might be adopted on a wider scale. First Nations peoples are famous for communication with other species, and it may not be too late to do so with Twelve Step adherents. The "horse whisperer" Monty Roberts learned his trade from First Nations peoples of the southwestern United States,⁴⁸⁶ and writes of his ability to communicate with horses, and of parallels in communication between horses and people. Al-Anon's book *Having Had a Spiritual Awakening* . . . talks of communication with other animals.⁴⁸⁷ And Ted Andrews has done wonderful work in his books, *Animal-Speak* ⁴⁸⁸ and *Animal-Wise*.⁴⁸⁹

Other Considerations in Group Process

First Nations circles have an Elder (or an experienced member if an Elder is not available) to provide leadership, depending on the purpose of the gathering. These circles might be, e.g., for healing, sharing, or teaching. Twelve Step groups ideally have leaders with six months to a year's experience who rotate in positions as "trusted servants" or who might be speakers. In later years, some call themselves "oldtimers," Twelve Step Elders. Twelve Step groups generally devote their meetings to discussion, to an experienced speaker or speakers, or to newcomers. In "outside"

---

484 This critical subject is developed further in succeeding chapters.
485 Al-Anon Family Groups, *One Day at a Time in Al-Anon*, 2000, 141.
486 Monty Roberts, *Horse Sense for People*, (Toronto: Alfred A. Knopf Canada, 2001), 40.
487 Al-Anon Family Groups, *Having Had a Spiritual Awakening* . . ., 25, 29.
488 Ted Andrews, *Animal-Speak: The Spiritual & Magical Powers of Creatures Great and Small*, (St. Paul, MN: Llewellyn Publications, 1998).
489 _____*Animal-Wise: The Spirit Language and Signs of Nature*, (Jackson, TN: Dragonhawk Publishing, 2001).

business and other meetings - among "our affairs" too - Twelve Step principles can be put into practice.[490]

Closed Twelve Step meetings are for members only, while open meetings are available to anyone interested in attending, for instance, members of the clergy, health care professionals, police officers, or students. Alcoholics Anonymous now includes open discussion meetings in its repertoire. Similarly, as noted, many First Nations circles now include those of non-First Nations heritage. This inclusiveness is about balance, openness, and recognition of the benefits of sharing in a shared humanity.

Spirituality, a sense of the sacred, is strong in First Nations circles. This includes "smudging" (cleansing) ceremonies with the burning of sacred herbs, sage, sweet grass, tobacco, and cedar; giving thanks to the Creator through prayer; the lighting of candles; the presence of sacred articles; teaching; the passing of the sacred talking stick, stone, or feather to denote the speaker; the smoking of the pipe; drumming, dancing, singing, and feasting.[491] In comparison, Twelve Step HPMA groups recognize the authority of Higher Powers in the Steps and Traditions, in celebration, prayer, literature, slogans, banners, and fellowship, all helping to make humility a virtue and spirituality a reality.

Twelve Step groups have set readings of preambles, the Twelve Steps, Twelve Traditions, the Serenity Prayer, the Responsibility Pledge, closings, and whatever else the group has decided upon with the guidance provided by the Traditions. Experience suggests that, over time, members may "tune out" and do not always listen to the important words that are being read. In Twelve Step meetings there is sometimes a tendency to

---

490 See William Hammond III, ed., *12 Step Wisdom at Work*, (Dover, NH: Kogan Page U.S., 2001).

491 Other First Nations ceremonies include presentations of the eagle feather, fasting, naming (when one acquires a spiritual name), sweat lodge, medicine wheel, and sundance ceremonies, the shaking tent, the potlatch, and those specifically for women or for men.

On Evolution

be late, or to leave early, "breaking circle," interrupting speakers and producing noisy disturbances.

Were it not for brief off-the-cuff remarks and spontaneous jokes strewn briefly about by readers, these set parts of Twelve Step group meetings might become tiresome. When these readings are for some reason unavailable, there may be some awkward stumbling about, but the job gets done. First Nations circles do not have a written text, and people who open and close meetings speak from their hearts and minds, with thoughtful creativity, and sometimes with the hesitation or humour that reminds one of Twelve Step meeting openings and closings.

Twelve Step group members often have about five minutes to speak in discussion meetings. First Nations talking circles have a more flexible approach and members may speak for fifteen minutes or more; this is unusual when there are time limits. Elders may share their wisdom in a teaching circle without being authoritarian. In both First Nations circles and Twelve Step groups, there is also flexibility, for instance, in a notable acceptance of those who stray from the suggested guidelines. The participation of less healthy people can still be of help to others: *all* have something to learn - *and* to teach.

Sponsorship is almost essential to the functioning of Twelve Step societies, especially for newcomers who may need extra support in learning about a different and better way of living. First Nations societies have a parallel system of support provided by Elders. Members sometimes speak of their Elders or sponsors during the course of First Nations and Twelve Step meetings.

The language of Twelve Step meetings sometimes creeps into First Nations circles, partly since many are for healing, as Twelve Step groups are themselves. Individuals in First Nations circles sometimes refer to their Twelve Step programs. These similarities across cultural lines are a clear recognition of the existence of similar approaches, and of the warmth and equality available within HPMA.

First Nations circles and Twelve Step groups are not without problems. The influence of outside, authoritarian, hierarchical society is very powerful, and in spite of traditions that place authority in Higher Powers, human authority sometimes enters the circle or group.[492] Dominance is an expression of it and is a problem from time to time.

One Elder spoke of the need to revive First Nations traditions, most of which had been forbidden until the past forty or fifty years.[493] The Elder concluded that First Nations circles had lost some effectiveness, due largely to the invasion of outside culture. With time an important commodity in today's world, sharing in First Nations circles is sometimes also curtailed to court the clock.

In both types of Higher Powered Mutual Aid, there is no expectation of perfection from individuals, or from the circles or groups themselves. Valuable lessons can be exchanged between First Nations and Twelve Step societies, one with the wisdom of age, the other with the vitality of youth. Groups in general might observe the principle of non-human authority more thoroughly, improve adherence to discussion guidelines, and ensure that meetings are spiritual. More circles and groups might be opened to wider membership, with newcomer meetings to introduce people to the HPMA way of life.

Issues of personal health led me to First Nations and Twelve Step HPMA circles and groups, and I have found that they are similar in key respects. But the wholesale destruction of First Nations cultures in the Americas has weakened the power and effectiveness of these older HPMA societies, and a now largely de-natured wider society is not yet a place where First Nations people can feel at home - in their own home.

Trust remains an issue. By 1600, the world's greatest "mortality" had been perpetrated by whites against the Indigenous peoples of the

---

492 In principle, A.A. is not authoritarian, but experienced members can offer advice to those who need/want it; this is allowable under A.A.'s Tradition Two, where God is the "ultimate authority."

493 Mainstream society is often still hostile toward First Nations peoples and their aspirations. Crises at Oka and Kahnawake in Quebec in 1990, at Ipperwash Provincial Park in Ontario in 1995, and in some reactions to the First Nations "Idle No More" movement of 2012, are only a few examples.

On Evolution

Americas; Wright uses a figure of perhaps 90 million dead.[494] Some would say that the genocide continues. According to a 2011 report by Amnesty International, the treatment of Indigenous people remains the leading human rights problem in wealthy Canada.[495] Whites must show themselves to be trustworthy when entering First Nations circles, long dominated by the destructive illnesses of Euro-American-Canadian cultures.

The Twelve Step pioneers knew, from their own trial-and-error experience, and that of others, that individuals recovering from the disease of alcoholism would benefit from following a somewhat structured approach. Hence, the evolution of the suggested Steps, Traditions, and Concepts,[496] and the benign anarchy of Twelve Step HPMA groups underlying their democratic societies. First Nations HPMA probably evolved over time, too, in an ecological, spiritual way of life, a natural state of being marked by serenity and individuality within the common welfare. It was no doubt always rooted in the great trial-and-error experience of living wisely with oneself, with others, and with all Creation.[497]

In this chapter, we have touched upon some remarkable similarities between First Nations and Twelve Step Higher Powered Mutual Aid. These make it plausible that Alcoholics Anonymous can be described as "clan-like."[498] HPMA is a natural, moral, healthy way of leading our lives, and the parallel successes of the two forms of HPMA give us hope that a brighter future may come our way.

---

494 Ronald Wright, *Stolen Continents: The "New World" through Indian Eyes since 1492*, 14

495 Amnesty International (Canada). *Getting Back on the "Rights Track": A Human Rights Agenda for Canada*, Ottawa, March 2011.

496 The responsible curious will find information on the Twelve Concepts in the Conference Approved Literature of A.A. and Al-Anon.

497 In his discussion of the move from gathering and hunting to agricultural societies, Campbell describes a possible suppression of individuality in favour of the interests of the common welfare. (Joseph Campbell, *The Masks of God: Primitive Mythology*, [New York: Penguin Books, 1969], 240).

498 A.J. Trevino, "Alcoholics Anonymous as Durkheimian Religion" *Research in the Social Scientific Study of Religion*, 4 (1992), 183-208.

There seems, however, to be reason for less optimism now. Current challenges - e.g., environmental ones - have grown to a far greater magnitude, in tandem with the inexorably rising population of the globe. Some think that it is already too late to turn things around. Still, a melding of ideas from First Nations and Twelve Step HPMA societies could play a crucial part in the healing of our "global village," specifically with a new and improved ecological morality, one that covers all the bases. Twelve Step programs emphasize improving morality; as noted, 7 of the Twelve Steps deal with the subject.

In view of the great need for alternative ways of living together peacefully and ecologically in the world, in humble and common cause, we would benefit immensely, I suggest and submit, from an application of Higher Powered Mutual Aid principles to individual, familial, communal, national, and world affairs, all amid the common welfare.

CHAPTER 6

Further along the Way Ahead

He who knows others is learned. He who knows himself is wise.

LAO TZE

IN THIS CHAPTER, we examine more of the tools that members can use within the common welfare of their First Nations and Twelve Step HPMA circles/groups and societies. I begin with details about the "individual in relation," a vital concept that I have found useful in fashioning my own recovery program, and in the fashioning of this book. The "individual in relation" is an expression of the fact that we are not isolated individuals, nor were we ever, but are related together in the web of all God's Creation.

The "individual" is a mere phantom, much better described as the "individual in relation" who has unbreakable interrelations with humanity, other societies, and with the cosmos. In the end, try as we might, the circle cannot be broken: human beings and others have always been, are now, and will always be "individuals in relation."

This concept is partly about choices with which we have been gifted as free individuals within the common welfare. It is precisely here that the relatively new Twelve Step societies offer us the most potential to change ourselves and change the world, for the better, possibly in partnership with much older societies of First Nations peoples.[499]

---

[499] Graveline explains the "self-in-relation" among First Nations (F.J. Graveline, *Circle Works: Transforming Eurocentric Consciousness*, (Halifax, NS: Fernwood Publishing, 1998).

James G. Duncan

While central importance is attached to the "individual in relation," it is considered to be self-evident because it is defined by the fact that we and other beings are always "in relation" to the rest of Creation, making choices or decisions that benefit ourselves and/or others. In stable, moral societies of all kinds, the individual good and the common good are mostly always served together, whenever choices or decisions are made by an "individual in relation" or by a group of "individuals in relation."[500]

Perhaps hidden from us by fear and pain, leading to illusion and/or delusion, and by a misleading and divisive "individual as separate," the "individual in relation" may sometimes be put off with a back wave of the hand, or at others taken for granted. In addition, understanding of the "individual in relation" may be omitted by lack of time for thoughtful exercise. Some may be too busy, or brainwashed, to notice the truth.

The latter part of this chapter looks at the Twelfth Step specifically, and how we might consider its implications more thoroughly in Twelve Step societies, as "individuals in relation" who wish to practice the principles of the program in *all* their affairs. The intention here is also to put into practice the Twelve Step slogan, "Keep It Simple." This may help us to untie the unnecessarily complex intellectual knots that bind our feet; walking the walk is not as difficult as it is sometimes made out to be.[501]

---

[500] The irresponsible farmer of Garret Hardin's understanding can become self-responsible, instead, if s(he) does not exceed the share of common land allotted for the grazing of her/his cattle.

[501] Gregory Bateson complicates his analysis of A.A. and falls into error. He says that Bill W. was diagnosed with alcoholism and stopped drinking in 1939; that year was 1934. Bateson also mistakes "governance" for one of A.A.'s aims and adds that A.A.'s Twelve Traditions are secular, failing to take into account the pivotal derivation of spiritual authority in A.A.'s Tradition Two. (Gregory Bateson, *Steps to an Ecology of Mind: Collected Essays in Anthropology, Psychiatry, Evolution, and Epistemology* [San Francisco: Chandler Publishing Co., 1972], 331-333).

On Evolution

Chewing on Ideas

At a friends' dinner table many years ago my playful father had me look the other way, only to place a tough portion of meat on my plate. The joke was on me and I felt it from others' smiles, if only faintly at the time. But as a member of the younger set, I could chew with sharp teeth and swallow more easily while cleaning my plate. I suggest that those of any age, including the longer of tooth, may wish to keep their teeth sharp for chewing on ideas in the world today. Democracy is a priceless commodity, requiring our attention, commitment, vigilance, and even love.

In general, we are in the habit of swallowing ideas before chewing adequately, and global mental/emotional/spiritual/physical indigestion, perhaps even constipation, has set in. Says Woodman, "...the kind of addiction our society is falling into [is] gobbling massive doses of undigested values and ideas."[502] It is important in particular, to examine - and to examine closely - the views of those who have influenced history the most, I need not remind the reader, that has brought us to the present where the Golden Rule has taken on a new definition, the mean: "He who has the gold makes the rules."

One of our great challenges, then, is chewing on ideas more thoroughly before swallowing. Edible ideas may then contribute to healthy digestion and incorporation in the body, mind, and spirit. Moreover, ideas need not be discarded completely to any "dustbin of history," nor can they be; even inedible ideas remain with us. Upon further reflection, they might be incorporated later, modified to suit, or chalked up under lessons learned. The world is still complex and varied enough.

Chewing on ideas more thoroughly stems from my curiosity, my skepticism, maybe my perfectionism, and from the need to create a safe environment for myself. It is rooted in the belief that each person can have unique opinions. Hearing others' ideas, for instance, in HPMA circles or groups, helps to sharpen my ear to the voice of the other, and to that of my Higher Power.

---

502 Marion Woodman in Nancy Ryley, *The Forsaken Garden*, 1998, 66.

But I was able to take a more thorough approach to this after I came to understand that only without fear can one listen and speak well enough. The deterioration of "individuals in relation," families, communities, nations and the planet - founded in the shifting sands of poor or blatantly wrong ideas and policies, incomplete or false beliefs, and biased histories - is directly and proportionally related to our distance from HPMA.

Armed with this understanding, I have railed ever more surely against the notion that competition is all in evolution. A culture reared on Darwinian theory is summed up nicely in the phrase, "Kick A__," popular with some in the T-shirt set; I wondered if "My" was omitted between the two unfortunate words. While such an expression may be intended to be funny, actually kicking someone is a violent act. I do have a sense of humour, on display sometimes, but not at the expense of important truths.

Bumper stickers can be humorous, too, and they can give one the warm impression, while burning gas in traffic jams, that progress is being made. The bumper sticker sentence "Practice Random Acts of Kindness" is read more than it is practiced. Moreover, one must question any proposal to randomize kindness, whether its intention is humorous or serious. Random kindness suggests that some legitimate unkindness could, or should, remain in one's life. The sentence lacks sincere purpose in warm compassion, or thorough understanding that kind *relationships* are the means to a better world, not a blithe, random kindness. Shouldn't kindness be extended to all in everyday life, for oneself and/or for others, in every aspect of our lives?[503]

At the first Al-Anon meeting I attended, the positive, attractive parts of the program at the newcomers' table were edible, outweighing the negative.[504] As Al-Anon suggested from the outset, the useful slogans,

---

[503] Another pet peeve is the arrogant notion that "anything is possible." Well then I might say in response to its proponent some moonlit evening, "Jump to the moon; take a run at it if you wish."

[504] I asked so many questions of the Al-Anon newcomers' group leader at my first meeting that she later said she thought I'd never return. I wanted all the answers before I knew the questions.

On Evolution

"take what you like and leave the rest," and "fake it 'til you make it" kept me coming back. But simply being in agreement with any and all is only codependency, not a moral, natural "individuality in relation" that promotes independence of thought, feeling, and action, all within the common welfare.

The most important case I have used in considering the work of others has been my treatment of the ideas of Peter Kropotkin and Charles Darwin on the role of cooperation and competition in natural selection. Some might say that the questioning of intellectual gods like Darwin is heretical or idiotic.[505] The old dictum "fools rush in where angels fear to tread" might capture their sentiments well. Were my questions about Darwin's ideas not supported by others, I might not have much of a leg to stand on. This point remains: again, considering the crisis that we are presently in, and have been in for some time, it is important to question past ideas simply because they have gotten us to the present, abounding as it is with so many serious problems.

A reading of Darwin's work shows that he placed nearly all his eggs in one basket in evolution, the merciless, competitive "struggle for existence." He put forward the dreadfully mistaken notion that innate competitiveness is nearly all there is in evolution, and that, unlike competitiveness, cooperation, mutual aid, and sociability, are not innate. His followers are said to emphasize this foolishness - no other word will suffice - even more, contributing to ongoing ecological, economic, political, scientific, and social problems. The nest has not only been soiled, its fabric is shredded and far less supportive than is needed for the continuation of the wonders of life on Earth.

---

505 Jastrow says of Darwin's work that "...no other single achievement in science has produced an impact on such seismic scale." (Robert Jastrow, General Editor, *The Essential Darwin*, vii). To Thomson, Darwin "is one of the most intriguing intellectual figures of the nineteenth century, a man whose ideas have continued to dominate society, and particularly science." (Keith Thomson, *The Young Charles Darwin*, [New Haven, CT: Yale University Press, 2009], vii).

James G. Duncan

The message that mutual aid - writ large in HPMA - is a more important factor of evolution than competition has gone unheard amid the noise of self-serving bigotry, conflict, and misinformation. Godless competition has held sway to the detriment of the planet, and it is difficult to see our way through the cultural and climatic smog. Yet we still have the tools to cooperate in that effort, if only we take them up. We can make the effort to become conscious to the fact that we are ecological "in relation" cooperative beings and can then take a more humble place in the natural sphere of Creation here on Earth.

During the discussion that follows, I consider the views of many authors whose ideas have stimulated my own. Their books have proven valuable chiefly for providing a context for suggestions about how HPMA can play a more effective role in transforming our lives and the world. My purpose here is to focus on further practical application of ecological HPMA principles to help us become freer, more moral, more self-regulatory, and more participatory within the common welfare of our democratic societies.

The Individual in Relation

No man is an island entire of itself; every man is a piece of the continent, a part of the main.

JOHN DONNE

The phrase "individual in relation" is a social and ecological construct used to describe the relations of the individual to others, to families, to communities, to nations, and to the Earth itself, in short, to all animate and inanimate Creation. In this section, the "individual in relation" and the practical, natural choice making that we share with all living entities, benefiting ourselves and others, are generalized as principles of life, truths, or facts describing life on Earth. This is a healthy, necessary way of seeing things, and of being, the way things must be if we are fit enough to survive.

On Evolution

In a First Nations and Twelve Step HPMA context, individuality is expressed within the common welfare of the circle or group, with choices or decisions made on behalf of ourselves and others. Without the safety provided first by circles/groups, our evolutionary cradle, fear will inevitably enter to dampen or shackle the creativity of even the most committed "individual," weakening or negating "individual in relation" purpose, and making it worse for families, communities, countries, and the globe.

Many if not most descriptions of the individual and common welfare do not account for the unity of the "individual in relation" and the commons. Murphy emphasizes the importance of the individual, the need to share personal experience,[506] and to first take care of our individual health.[507] Unfortunately, this leads him to focus on the notion that the individual can somehow be separate from others, and his analysis suffers. Says Murphy, "The concept of the 'individual' is at the crux of all ideologies and theologies, the root of all social systems, the critical issue of virtually all battles."[508]

Here, interrelations with others are not described as a critical, factual component of the individual and the individual's well-being, and societal links are only incidental to Murphy's discussion.[509] Though it may be far from his intent, he writes like a Darwinist champion of the "individual," an individual who he seems to think was separate from community at one time. He says, mistakenly, "One of the first evolutionary acts of humankind was the creation of communities and societies..."[510]

Human beings and all other life forms are innately social and cooperative, and always were, Darwinian theory notwithstanding. It is the purest fiction that it took a conscious act to create communities or societies. Diamond emphasizes the *unconscious* nature of decision-making in the move from gathering and hunting societies, to sedentary ones based

---

506 Brian Murphy, *Transforming Ourselves, Transforming the World: An Open Conspiracy for Social Change*, (New York: Zed Books, 1999), 5-6.
507 Ibid, 15-28, 134.
508 Ibid, 53.
509 Ibid, 62–64.
510 Ibid, 62.

on agriculture.[511] The individual is born into pre-existing community, if not society. What are families if not small communities, the building blocks for larger ones, such as the extended family?

Furthermore, these existing communities were always complex and ecological. From the beginning, communities have been composed of other living and non-living entities. The evidence is that we are relative newcomers to evolution, and communities were made up of all sorts of others to welcome us. Prey then, too, we survived to become part of these living communities, and did our share in building on the best agricultural land. Urbanization, so dominant today, has followed agriculture and industrialization.

In attempting to describe the individual's responsibility to promote the common welfare, Murphy operates on shaky ground. He says, "Our first step is not to decide whether our action will result in a new world, but, rather, to decide that our consciousness must result in action *regardless*, whatever the eventual outcome."[512] A proposal to take action "*regardless*, whatever the eventual outcome" is reckless, founded in frustration it appears, rather than cool calculation of the complicated factors of our interdependent existence.

Searching for the "structures through which basic change can be effected,"[513] Murphy is stymied by a *status quo* that prevents change. Yet, in his selection of ideas for such structures, Murphy completely omits First Nations and Twelve Step societies and methods from consideration, including in his extensive remarks about health.[514]

Swinging round the other way, Diarmuid O'Murchu de-individualizes the "individual in relation." For him, there is very nearly nothing but relatedness, and the individual is lost almost completely.[515] He writes,

---

511 Jared Diamond, *Guns, Germs, and Steel: The Fates of Human Societies*, (New York: W.W. Norton & Co., Inc., 1999), 104-13.
512 Brian Murphy, 1999, 10.
513 Ibid, 13.
514 Ibid, 15–23.
515 Diarmuid O'Murchu, *Religion in Exile: A Spiritual Homecoming* (New York: The Crossroad Publishing Company, 2000), e.g., 91, 115, 123, 137–38, 148, 152, 156, 169.

On Evolution

". . . our fundamental *raison d'être* is about relationality."[516] In his ninth chapter, "Reconnecting with Our Relational Selves," he says that, ". . . each of us is the sum of his/her relationships."[517] Here, O'Murchu denies the individuality contained within the "individual in relation."

O'Murchu says much about relating to others, to the Earth, and to the Universe, but there is no recognition of the "individual in relation" or of individual freedom and purpose within the common welfare.[518] In getting at his central and worthwhile ideas about reconnecting with the Earth, he says surprisingly little about the importance of mutual aid and cooperation, and the great evolutionary ideas of Darwin, Kropotkin, and others, are only incidental to his discussion.[519]

Murphy and O'Murchu start at and maintain opposing positions, perhaps a result of the former's secular and the latter's spiritual perspectives. But individuality and relatedness are not opposed at all. In fact, they are integral parts of the "individual in relation," supporting not opposing each other. The "individual in relation" contains the potential for both the individual good and the common welfare. The individual always was, and will always be, in natural relation to the rest of Creation. As in First Nations and Twelve Step HPMA practice, this is a relationship where:

- Contributions from each individual are needed to maintain the welfare of the circle/group.
- The individual has freedom to express unique purpose and creativity within circles/groups that are, in turn, safe and supportive of each individual.

---

516 Ibid, 91.

517 Ibid, 146.

518 Somewhat like O'Murchu, Saul seems to agree that our principal property is social. (John Ralston Saul, *On Equilibrium*, [Toronto: Viking, 2001], 8–9). All history and evolution shows that we are innately individual and social, together.

519 O'Murchu says that "...*cooperation* and not *competition* was an overriding principle [of early human evolution]." (Diarmuid O'Murchu, 179). Rather, it has always been thus, yesterday, today, and tomorrow.

It is worth restating that the individual always was, and will always be, born into pre-existing society. Societies had to exist for individuals to emerge - to be born, play,[520] work, and thrive - were it a family, a tribe, or other group in relation to its environment. Kropotkin observed in *Mutual Aid*, that "Society has *not* been created by man; it is anterior to [precedes] man."[521] This is not complicated stuff. But it does make for a confusing - and vicious - world when the "individual in relation" is described as a separate individual, when the individual good is placed ahead of the common welfare, or when it is understood that a competitive, "rugged" individualism will, somehow, yield the common good.

It is sometimes said that we humans are our own greatest enemies, and one wonders if that assertion is not true. Our great failure has been our inability to recognize and act upon the natural truths that result from the expression of individual freedom and purpose within the common welfare, exemplified in the practices of First Nations and Twelve Step Higher Powered Mutual Aid societies. The failure to see ourselves always and in all our affairs as "individuals in relation" to the common welfare, and to all Creation, has led humanity to ruin at the brink of disaster.

Natural Ambiguity as Truth

Helping us to make sense of the "individual in relation" to the common welfare, and its absolute importance in the world, we can conduct a searching and fearless examination of truth and fact, something also not as difficult as it might appear.

Truth and fact have slipped through our fingers and minds, a continuing difficulty stemming from our inability to recognize, or remember, that the group's "individual in relation" - not the fictitious "individual" - is a basic ecological construct. And stemming from this, natural ambiguity is

---

520 Peter Kropotkin, *Mutual Aid*, 54. Kropotkin says that all creatures like to play. More than once I have observed a crow and a black squirrel playing and teasing one another in a park near my home.

521 Peter Kropotkin, *Mutual Aid*, 54, note 1.

On Evolution

a fundamental truth (or fact) of our existence. Notably, dictionaries treat truth and fact as synonyms, chiefly in their roles as tests of reality.

One of the most simple and basic truths that we have the privilege of knowing is natural ambiguity. Its "naturalness" is the reality of life itself: all living things are individual and social/ecological, together, making choices and taking action for the benefit of oneself and for the benefit of others in the common welfare.

Natural ambiguity is all around and within us, created specifically by our existence as unique individuals *and* social, ecological beings. This ambiguity starts when we are born as "individuals in relation" to our families. We also have different sides: we are bipedal and bimanal, our brains are hemispheric, we think and we feel, etc., but only and always in relation to other parts of ourselves, and to others, within the common welfare.

The truth about ourselves is defined by natural ambiguity: it is true (or factual) that we are characterized by natural ambiguity. Such ambiguity leads us to be able to make decisions and take action for ourselves and/or for others within the common welfare. Our inability to understand this or to accept natural ambiguity as immutable truth (or fact), and to work with it to make the proper imperfect choices for ourselves, and in relation to the other, is characteristic of our age. This renders us unable to see that there are other real, solid, humanly knowable truths in the world, and unable to make consistent, healthy choices and decisions for the benefit of ourselves, others, and the Earth.

We will remain anxious, fearful, and confused, and serenity and happiness will elude us, if we cannot recognize and work effectively and safely with the natural, foundational ambiguity of life. Confusion and frustration over this important matter can lead us to run around in circles, rather than to sit still in circle or group, thinking, listening, and speaking, before acting, to make appropriate choices for ourselves.

Natural ambiguity is nurtured, or not, as the case may be. In dysfunctional homes and societies - e.g., those that are alcohol dependent, so much a part of recorded history - natural ambiguity can become distorted and distortions can be magnified. Choices between right and wrong, functional and dysfunctional, are blurred, to the degree that we may be unable to tell the difference between them. Right thinking, feeling, and acting, one day, for oneself in relation to the common welfare, can be wrong the next. Role models say one thing and do another, and children are left without adequate guidance from adults who might declare, "Do what I say, not what I do."

Through the practice of such distorted ambiguity, say, over 30 years or 3,000, we deftly change our actions, feelings, thoughts, and morals as circumstances present themselves to us in our daily lives. We have subsequently become slaves to, rather than masters of, *ambivalence*, taking refuge in tolerance of unacceptable behaviour from ourselves or from others. Such enslavement presents itself, as noted, as an inability to make consistent, moral, healthy choices and decisions for our individual benefit and for the benefit of others in the common welfare.

Ambivalence is different from ambiguity, and the difference can illustrate a way forward. The dictionary defines ambivalence as a noun meaning, "the coexistence in one person's mind of opposing feelings… in a single context." When opposing feelings coexist in a person's mind, it is difficult to make decisions. If one cannot decide whether to toss an empty pop can out the window of a car, or to keep it for recycling later, one is caught in a dilemma. If the opposing feelings continue, a choice is not made, and paralysis or confusion results, with the pop can held ready in one's lap, awaiting action or not.

Ambiguity, moreover, is a noun defined as a "double meaning," or something "able to be interpreted in more than one way." Easier choices for a healthy "individual in relation" are offered by a clear "double meaning." An ambiguity is something "able to be interpreted in more than one way," presenting real choices for oneself or for the other in relation to oneself.

Here, the actual tossing of the pop can out of a car window means that remaining ambivalence about the matter has been transformed into an unambiguous decision. However, such a decision may only be temporary. The next time around, ambivalence may rear its head again, with its attendant confusion or paralysis. If and when an unambiguous decision is made once more, a new pop can may be kept at one's feet for recycling. If this pattern continues, confusion does reign, not consistent, moral, healthy decision making for oneself and for the roadside.

Driving this home a little more, ambivalence is defined as a feeling, while ambiguity is defined as a meaning or an interpretation. The truth of our healthy, ambiguous existence as individual and social/ecological beings, together, is an internal and external truth or fact, represented by real, meaningful choices. This truth or fact cannot be sustained forever by unhealthy feelings of paralyzing ambivalence that, in the end, produces a toleration of *all* choices. Thus, one can use solid, natural, factual ambiguity to make clear, consistent, healthy, and moral decisions, either for oneself or for others in relation to oneself, all within the common welfare.

Discussion

In order to take effective action in our lives, we must be able to decide what is true (factual) and what is not true (false). Fortunately, it is plain to see, if we scratch the surface just a little, that there are simple, natural truths that can guide us in the conduct of our daily lives as we pursue our individuality in relation to the common welfare. The truth is "in there," in our hearts and minds, as well as "out there," in the world around. If we are to make sense of anything at all, we must keep in mind that the foundational truth begins with the natural, social, ecological, ambiguous "individual in relation." Unless we get this foundation right - and we haven't - the economic, moral, political, scientific, and social edifices that we construct will naturally fall away - and they are.

To many, the subject of truth seems impossible to pursue. Only the deep thought of philosophers, scientists, or theologians can penetrate it, we think. Or smashing protons and decoding genes can lead us to the truth, to "the language of God," in the words of one boastful scientist. I suggest that we need not rely on others for truth. Rather, it is discoverable by all. We have forgotten, or never learned, that there are many simple, natural truths, or facts, that sustain us all as "individuals in relation." Re-establishing a firm foundation with such natural truths (or facts), we might venture forth again with the confidence of the imperfect, very human beings that we are.

Murphy feels that the pursuit of truth is a pointless one since each person's truth might be different.[522] He is, perhaps, confusing truth with purpose: each person's purpose is different because each person is unique. Promisingly, Murphy says that we must "accept the inherent ambiguity of human existence."[523] However, he does not follow this through to its ecological conclusion, namely that the natural ambiguity of the "individual in relation" is *itself* the foundational truth. Murphy's is a false start, then, leading him to say that "…truth has no validity…" and that "…we can *never know* whether what we hold to be knowable fact is, in reality, true."[524] Nonsense.

In his chapter, "The Missing Link,"[525] Murphy begins with several statements, beginning with, "First and foremost, human beings are creatures." Brief explanation follows to the next statement, that "Second, human beings are living creatures," and to the next, "Third, we are intelligent creatures." Here, I had to admit that there may be some validity to his idea that truths may be different for some.

For instance, are human beings all that intelligent? Population growth has no end. Addictions are rampant. Mental ill-health is near the norm. People go hungry and homeless amid plenty. Deadly wars and conflicts continue. Garbage and toxins are strewn about the world, filling the air, land, and water. We eliminate "competing" species in an unremitting crusade that is actually suicidal,[526] bent as it is on destroying the social and ecological systems that support us. So, numerous, yes, dominant, yes, creative, yes, but intelligent?

---

522 Brian Murphy, *Transforming Ourselves, Transforming the World*, 1999, 31–32.
523 Ibid, 32.
524 Ibid, 58-59.
525 Ibid, 43–51.
526 Carl Jung recognized our suicidal course. (Meredith Sabini, ed., *The Earth has a Soul*, 22, 126). Ronald Wright agreed in his *A Short History of Progress*, 121, 131.

On Evolution

David Suzuki quotes Paul Hawken as saying "...we are not very bright."[527] Suzuki himself says that fouling the nest through widespread pollution is "hard to understand,"[528] and that humans are not good with "three-dimensional reality."[529] According to Wright, waste is our "most immediate threat."[530] Hawken supports this, stating, "...ultimate threats to human welfare" may be environmental ones.[531] Wright concludes that *Homo sapiens* ("wise man") may be "clever but seldom wise."[532]

We appear to be successful, but, as ever, appearances are deceptive, and with growing problems that somehow surprise us, we ignore them, blame others, or whine that we are "victims of our own success." More exactly, we are victims of a false understanding of ourselves as being separate from the rest of Creation, as being apart from an interconnected whole, and as being unrelated individuals and communities.

With such false understandings, we can only be successful at making victims of ourselves. Perhaps we are less intelligent than we think, and one wonders if the stronger current is something like self-hatred. Murphy's reference to our being living creatures may have a hollow ring not far down the road. I see more and more robotic people, gesticulating or driving, connected to cell phones or other technological marvels.[533]

---

527 David Suzuki et al, *Good News for a Change: Hope for a Troubled Planet*, (Toronto: Stoddart Publishing Company Ltd., 2002), 87.

528 Ibid, 320.

529 Ibid, 346–47.

530 Ronald Wright, *A Short History of Progress*, 7.

531 Paul Hawken, *The Ecology of Commerce: A Declaration of Sustainability*, (New York: HarperCollins, 1993), 202.

532 Ronald Wright, *A Short History of Progress*, 132.

533 I have been skeptical of some advances in technology, perhaps a neo-Luddite. However, one evening while answering the phones for Al-Anon, I received a call from a newcomer seeking her first meeting. She had gotten the Al-Anon number from a search on her Blackberry and used the same device to call me!

James G. Duncan

In his discussion of ambiguity and truth, Mark Kingwell ups the ante and repeatedly stubs his toe on the notion of "ultimate truth," borrowing heavily from the thought of French philosopher Michel de Montaigne; Kingwell says:

> Precisely because Montaigne believes no knowledge of ultimate truth is available to the human mind, he moves in strikingly modern fashion to adopt a position of ethical and political tolerance.[534]

Buying into Montaigne's argument holus-bolus, Kingwell says that, since we cannot find ultimate truth, we might or must therefore adopt ethical and political tolerance. He is thus led to tolerance by the impossibility of finding ultimate truth. A key observation is this: the futile search for ultimate truth provides a weak foundation for an adequate factual and moral tolerance. A key question is this: since there are many *humanly knowable* truths, or facts, why not build a solid base for ecological, economic, moral, scientific, social, and political decision-making and structures on those? Here I differ heartily with people who say that truth is unfathomable.

Why do we long for ultimate truth in the first place? It would be easier looking for a needle in a haystack on a cloudy moonless night without a flashlight or magnet. The search for ultimate truth may be entertaining but will always elude us. Even Einstein felt that the Universe is "only partly knowable."[535] Lovelock discussed this in 2006.[536] The pursuit of ultimate truth is only a red herring, a straw man devised of our frustration at being human and not God. While ultimate truth is not available to us, this does not mean that we can forget, deny, or trample upon the truths that *are* available.

Critically, the practice of tolerance should stem, not from our understandable failure to find ultimate truth, but rather from our success

---

534 Mark Kingwell, *The World We Want: Virtue, Vice, and the Good Citizen*, (Toronto: Viking, 2000), 112.

535 Larry Dossey, *Recovering the Soul: A Scientific and Spiritual Search*, (New York: Bantam Books, 1989), 146.

536 James Lovelock, *The Revenge of Gaia*, (London, UK: Allen Lane, 2006), 138.

at being able to understand and work with humanly knowable truth. The tolerance that we have today is rooted in an "anything goes" culture, where we are unable to distinguish humanly knowable truth, or fact, from fiction, right from wrong. We begin to tolerate and then practice habitual permissiveness amid social and ecological immorality. We can be ecologically moral with voluntary simplicity, a life away from over-consumption and towards lessening our enormous ecological footprint on the Earth.

The habit of living in the head, rather than in the head and the heart together, leads to difficulties in identifying humanly knowable truths, such as the first basic truth of human existence, the natural, ambiguous "individual in relation." In spite of reservations about our intelligence, we can clearly identify and use many other humanly knowable truths. They are legion, presented to us by our place on Earth as cooperative, competitive, and social beings - "individuals in relation" to the common welfare and to all Creation.

In place of Kingwell's tolerance, I prefer acceptance. In Twelve Step fashion, individual and collective human limits[537] can be accepted in the knowable truths of our existence on Earth as "individuals in relation." Stemming from this, what may seem an imperfect truth or fact - natural ambiguity - is actually and unavoidably perfect for our existence as human beings who are, and will always be, "in relation" to Creation.

Humanly Knowable Truth

Penetrating so many secrets, we cease to believe
in the unknowable.
But there it sits, nevertheless, calmly licking its chops.

H.L. MENCKEN

---

537 "[W]e humans," observes Benyus, "regard limits as a universal dare, something to be overcome..." (Janine Benyus, *Biomimicry: Innovation Inspired by Nature*, 1997, 7).

James G. Duncan

Although we are not privy to ultimate truths, there are many humble truths that we can and do know from the perspective of our shared humanity and shared planet. These truths actually govern our existence and are knowable by all. They are general, moral, natural principles, if you will, defined by our interdependence and held solidly in place in this ecological place called Earth.

Yet we have tried, over the last millennia, to make fictitious, incorrect, unnatural "individual" *independence* work, rather than factual, natural "individual in relation" *interdependence*. Strict individualism is no more than an immoral myth. It is anti-ecological, anti-human, and anti-Earth. Except in the short term, geologically speaking, it cannot and does not work, never did, and never will.

Murphy speaks of the Earth as prison-like.[538] The Earth may feel like a prison, but in reality, we are only prisoners of our own fears and misconceptions. That the Earth is our only home is the overarching truth stemming from our "individuality in relation": each of us is in a natural, give-and-take, ecological relationship with the Earth. Should we continue to forget or ignore that the Earth is our only home, so far, She will no doubt survive the ordeal, rid herself of us, and find other children who like Her better. Nancy Ryley quotes Brad Hunter as saying, "It is almost as if Mother Earth recognizes the human species as a threat to her well-being, and her immune response is to cleanse herself of this parasite."[539] Carl Jung stated perceptively, "Nature *must not* win the game, but she *cannot* lose."[540]

Some of the following truths are self-evident; there may be those among us who say that, therefore, they are not worth talking about. It might also be said that, since self-evident truths, or facts, are not the product of deep thought, they can be ignored, as noted. The fact is that, to our everlasting peril, many of these truths *are* ignored, or honoured in the breach, in our thoughts, words, and actions. If these truths seem so

---

538 Brian Murphy, 1999, 136.
539 Nancy Ryley, 1998, 7.
540 Meredith Sabini, ed., 2008, 210.

obvious that we continue to ignore them, then ignore them we will. That is our choice.

This "Keep It Simple" approach to truth is also "Keep It Basic." If we concentrate on basic truths, and have these firmly planted in our hearts and minds once again, it will be easier to work from them as a firm foundation upon which to build, to heal ourselves and the planet. The following are some of the vital truths that flow from seeing ourselves as "individuals in relation" to Creation; citing them is not meant as condescension.

- It is true that individuals have always been and will always be "in relation" to others and to the common welfare. We are cooperative, social, competitive, and ecological creatures, and have a give and take relationship with each other, the Earth, and all Creation.

Our existence is therefore inseparable from the truths, or facts, that:
- The ground beneath our feet supports us, the valleys, the plains, and the mountains. With pollution, urban encroachment, and deforestation, habitat destruction is widespread, with resulting loss of foodland for children of every species.[541]
- The air we breathe and in which we spend our existence, supports us. Air is becoming increasingly polluted, adversely affecting vegetation, land, and water, too, and there is less and less vegetation to produce oxygen, to use carbon dioxide, and to cool the planet.
- The water we drink and bathe in, and the rivers, lakes, seas, and oceans in which we swim, and from which we draw liquid sustenance, support us. They, too, are polluted, killing people and other mammals, fish and birds, forcing us - imagine this even 20 years ago - to bottle and sell fresh water, the cleansing lifeblood

---

541 "Foodland" - farmland - needs to be protected, with farmers compensated fairly.

of the Earth.[542] With rising sea levels, islands, fertile coasts, cities, and other communities may disappear.

- Our families, communities, and countries support us. These are often dysfunctional, chiefly because we are losing the ability to recognize and practice the basic truth of our existence as "individual in relation" ecological, economic, political, and social creatures. The human invasion of the entire Earth,[543] immorally overpopulating and homogenizing it, is making it more and more difficult to simply live.
- Other living and non-living entities support us. We are not primarily in competition with other animals, land, vegetation, water, etc., for space. We are primarily in cooperation with them in mutual aid relationships.

First Nations and Twelve Step Higher Powered Mutual Aid principles rest on basic, humanly knowable truths such as are outlined above, respecting all of Creation, leaving ultimate truths in the capable hands of the Higher Powers. While pioneering has often taken on the meaning of physically going somewhere else (e.g., in space travel), the Twelve Step pioneers stayed put to deal with their own problems.

Their example is instructive. Like them, we must get our own house - home - in order first. We must begin with our individual selves "in relation" to the common welfare, with the critically important truths that have been available to us since time began.

Imperfection

Earlier I said I preferred the word acceptance to Kingwell's tolerance. Kingwell uses tolerance again, this time in relation to imperfection. Specifically, he says that in the distant past "the key to resolving and

---

542 In 2013, 1.2 of the 7 billion people in the world lacked clean drinking water. (*Ottawa Citizen*, p. A13, March 23, 2013).

543 As noted, Curtis Marean titles his *Scientific American* article, referring to human beings, as "The Most Invasive Species of All."

On Evolution

managing the deep conflicts of pluralistic politics was a willingness on the part of citizens to tolerate imperfect solutions."[544]

One of the definitions of tolerate is to "endure or permit, especially with forbearance," while forbearance means "patient self-control or tolerance." As Kingwell uses it, "tolerate" seems to have an element of frustration within it, perhaps associated with having to endure imperfection. Such a use of tolerate would be difficult to associate with the acceptance of the Serenity Prayer.

More recent history (i.e., now and in the past century) is a litany of deep conflicts, always looming, with less tolerance and no resolution in sight. Thus, I would definitely want to use acceptance in place of tolerance. That is, in Twelve Step fashion, I suggest that we might accept rather than simply tolerate imperfect solutions, faithfully and patiently working toward less - or more - imperfect solutions.

Using Kingwell's analogy of the key (above), the key to accepting imperfect solutions is knowing that, in working toward them, we have only the use of an imperfect key. This then becomes the imperfect key to the acceptance of imperfect solutions. But before we are able to use this key in an effective manner, we must step back to use three more separate keys, all of which are imperfect. These keys are:

- Self-respect, self-regulation, and self-responsibility through knowing and honouring our own abilities and imperfections, such as our ability to use an imperfect key.
- Respect for others in the common welfare, and for others' abilities and imperfections.
- Respect for the naturally ambiguous, ecological "in relation" foundational truth that keeps humanity together in humble and common cause.

This may sound complicated, but isn't. It should come as no surprise that all of these imperfect keys to acceptance can be found in Higher Powered Mutual Aid.

---

544 Mark Kingwell, *The World We Want*, 8.

James G. Duncan

Dialogue

Murphy writes of the importance of dialogue in several places,[545] and Suzuki observes the value of face-to-face dialogue in generating trust.[546] Kingwell speaks of the internalization of the virtues of dialogue in order to promote societal well-being.[547] In First Nations and Twelve Step HPMA, dialogue is carried out in safe, non-authoritarian circles and groups, where achievable objectives are sobriety, sanity, serenity, self-regulation, morality, and the sense of who we are in relation to others.

These virtues of dialogue are internalized as part of the healing process. Dialogue in First Nations circles and Twelve Step groups encourages "individuals in relation" to listen and speak without fear, in ways conducive to their own healing. But First Nations and Twelve Step HPMA dialogue also benefits circle/group effectiveness and cohesion. The manner of listening and sharing is consistent with the discussion guidelines referred to earlier, as the trust-building, fear-reducing, "sleepers of safety." Finally, such dialogue recognizes that the circle/group cannot survive without individual contributions to its welfare. Nor, of course, can the individual survive without the circle/group.

The outward expression of internalized virtues of dialogue is also a virtue of dialogue; this is the externalization of internalized virtues. Through equitable dialogue in First Nations circles and Twelve Step groups, we learn a sense of self and purpose in relation to the common welfare, and can then carry the message to others more effectively in all aspects of our lives. It is exactly this that is at the beginning of a politics that can help us heal ourselves before we do anything else. Only then can we begin to make sense of a planet in disarray, and engage in quiet, thriving dialogue as sober, creative, democratic, ecological, free, moral, responsible, sane, self-regulating, serene citizens, in our families, communities, and countries, the world over.[548]

---

545 Brian Murphy, 5–6, 18–19.
546 David Suzuki, 2002, 18.
547 Mark Kingwell, 8.
548 If people everywhere practiced this, over a relatively short time "peace would break out," as Karen Armstrong hopes in *Twelve Steps to a Compassionate Life*, 131.

A little later in his book, Kingwell claims that, "...as ever, seeing beyond ourselves is the only way to begin realizing the complicated project of emancipation."[549] Similarly, O'Murchu, emphasizing relatedness throughout his book, says, "It is no longer appropriate (in fact, never was) to begin with the individual and move outwards."[550]

As we have seen, such notions run quite opposite to the principles of Higher Powered Mutual Aid. In First Nations circles and Twelve Step groups, we learn, explicitly, that emancipation begins with our individual selves "in relation." Only by liberating ourselves and our creative potential within the common welfare of our circles and groups, can we effectively carry the message of hope.

Kingwell's "complicated project of emancipation" can be rendered child's play if emancipation begins with the natural "individual in relation," and if a critical mass of people begins to engage in its seeming self-indulgence. Young children know intuitively that we begin with ourselves but are often taught otherwise in anti-ecological, authoritarian, hierarchical human society. The choice to change to a focus on our "individual in relation" selves, first, still presents itself to us throughout the world.

Citizenship

Later, we review some of Bill W.'s writings about the restoration of citizenship through the Twelfth Step's "in all our affairs." He demonstrated a commitment to this, as did his wife, Lois W., and other Twelve Step founders. I believe that this restoration can be the product of an adequate recovery in HPMA circles and groups, as well as an enormous step forward for ailing "individuals in relation" to the ailing planet.

Kingwell discusses citizenship in the world today, saying optimistically that a new type of world citizenship "[tending] towards justice," will become available to us.[551] Observing that "reflection on the idea of

---

549 Mark Kingwell, 73.
550 Diarmuid O'Murchu, 169–70.
551 Mark Kingwell, viii–ix.

citizenship is in painfully short supply, whether from . . . *busyness* or from *knowingness*, the twin distracting deities of our day," Kingwell adds, ". . . these linked preoccupations are an attempt to sidestep the anxiety we actually feel about ourselves and our world."[552]

"Knowingness" is workable as a place to begin if it is first about knowing oneself, in the context of supportive circles/groups and communities. However, this does not appear to be what Kingwell is getting at. His "knowingness" may be an intellectual preoccupation, a kind of pouring knowledge into craniums available at many learning institutions.

A thoughtful First Nations man spoke at one University, shaking his head and declaring its existence "a complete mystery." Perplexed at the time, I have since thought that he might have been questioning how society creates centres of knowledge and metes it out as an expensive commodity.[553] Mainstream hierarchical institutions of all kinds may be monuments to fear, autonomically dissecting the self and the other, but unable to understand, explain, and practice the relatedness of the "individual in relation" to the common welfare and to Creation.

A great malady of the age, a tragedy in fact, is leaving unexplored the vast territory of the individual self "in relation." The irresponsible hiding away of the self "in relation" is the source of, or the reason for, much of the anxiety that we face. Going it alone, as "out laws," without HPMA in our lives, is literally sickening. Counteracting the anxiety with HPMA healing and recovery is the preoccupation that we must engage in, to reduce our anxieties and to increase our sobriety, sanity, and serenity, so that we may effectively carry the message to others in every aspect of our lives.

Kingwell writes of the need for belonging.[554] But belonging can first exist only with being. Thus, we might begin with the need for being, that is, simply accepting ourselves as *being* fully human, with all our strengths and weaknesses. Furthermore, we must know ourselves "in relation," in

---

552 Ibid, 4.

553 Quoting a California farmer, Benyus says that First Nations people "didn't need schools and churches - their whole world was one." (Janine Benyus, *Biomimicry*, 1997, 11).

554 Mark Kingwell, 4–5.

order to *be* ourselves, and to continually *become* ourselves, exploring our potential so that we can properly belong to ourselves and to others in circles or groups of citizens. Knowing oneself in relation to the commons through HPMA practice is the first step that we must take to return to healthy being, belonging, and becoming.

As First Nations and Twelve Step societies demonstrate, we define who we are as individuals first, in HPMA circles and groups to which we safely belong. Knowing ourselves, our "individual in relation" talents and flaws, in such circles/groups, and in all our affairs, is the basis of a strong personal foundation. Only then can we learn to put our meaningful, positive purpose(s) for individual living in relation to the common welfare, into the mutually beneficial practice of, to repeat, being, belonging, and becoming.

What to Change?

The greatest problem with many approaches to social engineering is that they do not begin definitively with changing the self "in relation." Murphy lauds the individual and leaves out relatedness. O'Murchu writes of relating but omits the individual. In the real world, they are not separate, never were, and never will be. Dass and Bush write about service in *Compassion in Action: Setting Out on the Path of Service*, but their chapter "Start Right Where You Are"[555] does not start where you are at all. In fact, the "you" is left out, and the assumption is made that the helper is able to give and give continuously.

Twelve Steppers know that endless giving is almost as sick as endless taking. People can be oriented toward others to the degree that self-care is forgotten.[556] In some places, this is called codependency, clinging to others and preoccupied with trying to change or control them in different ways. Pointing fingers at others and their plight seems easier than accepting

---

555 Ram Dass et al, *Compassion in Action: Setting Out on the Path of Service*, (New York: Bell Tower, 1992), 180–87.

556 Self-care can be forgotten when someone else is in danger that we can prevent. Altruism is a fact of life around the world, if we are true to our mutual aid natures.

ourselves, others, and all of our imperfect natures. In not looking, first, in the mirror at our own defects and qualities - if only to define ourselves as imperfect - we become part of the problem not the solution.

Of course, imperfections that are actually evil must be rooted out, either our own or that of others. The Twelve Step slogan "Live and Let Live" does not apply when evil has been, or is being, committed or planned. Evil is defined, I suggest, as anything that opposes or contradicts the ecological and moral application of HPMA societal principles. It is not alright to plan and carry out ethnic cleansing and terrorism. It is not alright when one tosses a pop can onto the roadside. Tracking down wrongdoing is accomplished in A.A.'s Step Four, with each person following the suggestion to make "a searching and fearless moral inventory."

We can learn the way to sobriety, sanity, and serenity by re-learning how to be truly human in HPMA. We can then end our habitual participation in a confusing, overly tolerant, nightmarish culture, by learning instead to habitually participate in the practice of the healthy, moral, spiritual HPMA principles of the "individual in relation" to the common welfare. This can be the natural foundation for a better world, built with adequate safety for ecological, moral, self-regulating, self-responsible, and social individuals, and adequate safety in and for ecological, moral, responsible, self-regulating, and social communities.

Knowing oneself is the starting point, always the first step that we must take if we want to succeed in recovery, for our benefit and for that of the planet. If we take that step, we can take the second, seeing, and seeing to it, that our individuality is always relational, where the protective common welfare comes first, so that the individual can be as free, healthy, and whole as possible. The Twelve Step movement has made an excellent beginning for the practice of a healing "individual in relation" morality. With enormous thanks to ancient First Nations peoples and to the Twelve Step pioneers, HPMA is already available to us, almost worldwide. As the thoughtful First Nations woman said, we might yet find the way back to the Garden that still exists.

CHAPTER 7

Twelfth Step work "in all our affairs"

Let our individual members heed the call to every field of human endeavor. Let them carry the experience and spirit of A.A. into all these affairs, for whatever good they may accomplish.[557]

BILL W.

THE VARIOUS THREADS of the book are brought together in the three remaining chapters to illustrate some practical ideas for the beginnings of a much larger Higher Powered Mutual Aid project. This takes us beyond the immediate healing and recovery work for "individuals in relation" that First Nations and Twelve Step HPMA societies are known for, into what Bill W. referred to as responsible "citizens of the world."[558] This invites us to address outside matters provided by, for example, climate change, democracy, ethnic cleansing, human rights, and peace. This chapter deals with the practice of HPMA principles "in all our affairs," the often neglected final words of the Twelve Steps.

It is not frequently noted, at Twelve Step meetings and elsewhere, that, from the very start, Bill W. and the pioneers developed an effective approach to the last part of Twelfth Step work, practicing Twelve Step

---

557 *Alcoholics Anonymous Comes of Age*, 232-33.
558 *Twelve Steps and Twelve Traditions*, 177; *As Bill Sees It*, 262; *The Language of the Heart*, 327.

principles in all facets of our lives.[559] In this context, we can examine some of the founders' early writings. To introduce the subject in an organized fashion, we can turn first to the Responsibility Pledge of A.A.

A.A. Twelfth Step work "in all our affairs"

The Alcoholics Anonymous Responsibility Pledge boldly presents the vision of the A.A. pioneers for transforming ourselves and the world. A.A. adopted the pledge as the "Responsibility Declaration" at its International Convention in Toronto in 1965. It places responsibility for A.A. Twelfth Step work squarely in the hands of its members, and reads:

> I am responsible.
> When anyone, anywhere, reaches out for help,
> I want the hand of A.A. always to be there.
> And for that: I am responsible.[560]

While priority is given to helping alcoholics, the Responsibility Declaration reaches beyond alcoholism to helping "anyone, anywhere." Early in my recovery, I attended an A.A. discussion meeting and was surprised to hear one member share his literal interpretation. Emphasizing "anyone, anywhere," he elaborated to indicate that this meant he had the responsibility to reach out to help others in every aspect of his life, not only those directly related to his recovery from alcoholism.

Later, I found that this courageous approach, bringing the program to life in all parts of one's existence, was not new. Bill W. himself spoke and wrote about how we might practice the principles of the program in all our affairs, committing ourselves to a change of morals and values to respect ourselves and all Creation. The A.A. literature has a large number of

---

559 When summarizing the Twelfth Step, Makela et al omit practicing the Twelve Step "principles in all our affairs." (Klaus Makela et al, *Alcoholics Anonymous as a Mutual-Help Movement*, 119).

560 Reprinted with the permission of A.A. World Services, Inc.

references to problems related to excessive drinking, and therefore to the need for broader, deeper change. We continue with *Alcoholics Anonymous*, "The Big Book."[561]

ALCOHOLICS ANONYMOUS (1939)

When A.A. was in its infancy in the mid to late 1930s, *Alcoholics Anonymous* was written mainly by Bill W. It is a recording of the experience of the pioneers and what worked and what didn't in the early Higher Powered Mutual Aid work of alcoholics and their families. Something worked, and by 2000, about 21 million copies of the first three editions had been distributed around the world,[562] translated into forty-three languages.[563] The fourth edition was published in 2001.

Twelfth Step work was a part of A.A. experience right from the start, and practicing the principles "in all our affairs" receives prominent early mention. In Chapter Two, "There Is a Solution," Bill W. wrote that putting a stop to drinking was only the beginning, and that a much more significant matter was putting Twelve Step principles into practice at home, at work, and in other facets of living together with others.[564]

A.A. numbers were low in 1939, only about 100 members, and the focus of Twelfth Step work was carrying the message and recruiting new prospects for the nascent A.A. In Chapter Seven, "Working with Others," the Big Book talks of carrying the vital message to other alcoholics.[565] The element of divine inspiration was important here, perhaps during that time especially when A.A. groped slowly ahead with the guidance of Higher Powers and, initially, the Oxford Group. Dedication and hard work were accompanied by spiritual awakenings, all key elements for Twelve Step HPMA from the outset, when A.A. also knew that "faith without works

---

561 This is a "fond title" given it by A.A. members. See *Alcoholics Anonymous,* 2001, book jacket.
562 Alcoholics Anonymous (The Big Book), 4th edition, 2001, book jacket.
563 Ibid, xxiii.
564 Ibid, 19.
565 Ibid, 89.

is dead."[566] Twelfth Step work never strayed from the principles of Higher Powered Mutual Aid, namely that alcoholics would benefit from reaching out to help themselves and others. Chapter Seven ends with prophetic understatement, with Bill W. hoping that A.A. might eventually help the public to realize the severity of alcoholism.[567]

Following the seventh chapter of the Big Book, passages of Bill W.'s writing stand out as examples of early thinking about how to practice the principles of the Twelve Step program in the many different aspects of life. Chapter Eight, "To Wives,"[568] and elsewhere, details social problems that can arise in marital relationships threatened by the family disease of alcoholism. It speaks of a scarcity of friends, disgusted employers, financial insecurity, and telling children that the alcoholic was just sick.[569]

Further, Chapter Eight says that some alcoholics are thoroughly incorrigible and that patience and understanding will not make a difference.[570] A little later, Bill W. expanded on this to say to spouses of alcoholics that there may be times when life must be started over again with a clean slate, made smoother with a spiritual approach.[571] Here, Bill W. seems to suggest that spouses might also adopt the spirituality of the A.A. way of life, predating the birth of the Al-Anon Family Groups in 1951.[572]

Chapter Nine, "The Family Afterward," relates to domestic life, with Bill W. making several worthwhile suggestions. First, instead of closing the door tightly on the past, trying to forget all its horrors, we can bring it out into the light of day in the knowledge that our experience might

---

566 Ebby T. related this to Bill W. in 1934. (*Alcoholics Anonymous*, [The Big Book], 14). The Bible says, "For as the body without spirit is dead, so faith without works is dead also." James 2:26.

567 *Alcoholics Anonymous* (The Big Book), 103.

568 In A.A. at the time, nearly all spouses of alcoholics were female.

569 *Alcoholics Anonymous* (The Big Book), 18, 105–106.

570 Ibid, 108.

571 Ibid, 114.

572 The 1939 Big Book has a pre-Al-Anon story, "An Alcoholic's Wife." (Alcoholics Anonymous, *Experience, Strength and Hope*, [New York: A.A. World Services, Inc. 2003], 128-29).

become a treasured possession that we can fruitfully share with others.[573] Second, as regards children, Bill W. suggested patience: give them time to get over their hurts, to see that the alcoholic is a new person. Children may trust again, and can take part in daily activities without hard feelings.[574] Chapter Ten, "To Employers," reaches out to business people, recognizing the profound effects of alcoholism in the workplace.

The "Personal Stories" section of the Big Book is important, as Bill W. said, a crucial method of making contact with those still outside A.A., but looking in.[575] In the 1939 first edition, 28 people shared stories of how their lives were affected by alcoholism and how they found, in A.A. and the Twelve Step way of life, the means and support to enable them to reach sobriety, sanity, and serenity.

In the 2001 fourth edition of *Alcoholics Anonymous*, expanded to 42 stories, alcoholics continued to speak of the harm done, not only to themselves, but also to their families and communities. Alcoholics' participation in A.A., as described therein, helped remedy the damage done to those affected by their drinking. Even by 1939, Bill W. and the pioneers understood that alcoholism creates serious problems in the workplace, in the family, in the community, and in all relationships. Additionally, they had already identified some effective means of dealing with these problems.

Twelve Steps and Twelve Traditions (1953)
Bill W.'s discussion of the Twelfth Step in *Twelve Steps and Twelve Traditions* remains its strongest interpretation to date.[576] Written by the renowned A.A. cofounder in the Cold War period, these words may be the first unmistakable sign that he and other pioneers thought that Alcoholics Anonymous might change the world. They are among the most powerful words Bill W. ever wrote.

---

573 *Alcoholics Anonymous* (The Big Book), 123-24.
574 Ibid, 134.
575 *Experience, Strength and Hope*, 2003, ix.
576 *Twelve Steps and Twelve Traditions*, 1953, 106–25.

After writing about the first part of the Twelfth Step, spiritual awakenings and carrying the message to others, Bill W. asks about the last words of the Twelve Steps, putting the principles into practice "in all our affairs." This last phrase of the Steps, he says, is compensation for all the hard work and sacrifice of working the A.A. program.[577] Two pages later, he elaborated, asking a series of rhetorical questions on the subject of "in all our affairs." Some of these are paraphrased below, as follows:

- Might we apply the program principles to our whole lives as much as we do to the part that we discover when we try to help other alcoholics achieve sobriety?
- Might we have the same respect and serenity in our sometimes dysfunctional families as we have in our A.A. groups?
- Might we carry the A.A. principles into our daily occupations?
- Might we bring new purpose and devotion to the religion of our choice?
- Might we come to terms with seeming failure or success?
- Might we accept poverty, sickness, loneliness, and bereavement with courage and serenity?
- Might we meet our responsibilities to the world as well?

With a resounding affirmative answer, and no equivocation, Bill W. replies that all of these are possible when practicing the principles of the Twelve Step program "in *all* our affairs."[578] Bill W. emphasized that "in all our affairs" originated with William James in *The Varieties of Religious Experience*.[579] *Varieties* was very influential in the Oxford Group, where demonstrating the principles in all aspects of life was an utter requirement, as Ebby T. explained to Bill W. in 1934.[580]

---

[577] Ibid, 109.
[578] *Twelve Steps and Twelve Traditions*, 111. Italics from *Twelve Steps and Twelve Traditions*.
[579] *The Language of the Heart*, 298.
[580] *Alcoholics Anonymous* (The Big Book), 14.

On Evolution

The family disease of alcoholism affected Bill W. from an early age. His near death at birth, his father's and paternal grandfather's heavy drinking, his mother's long illness, and his parental abandonment when he was 10 years old, have all been described at length.[581] Bill W.'s maternal grandparents took on the responsibility of raising him.

Many years later, Bill W. developed a great understanding of alcoholism, its sometimes traumatic effects, and the powerful goodness that is released in Twelve Step HPMA recovery work. Together with the experience of his drinking years, he was able to write about the disease of alcoholism from his own life, and about the need to deal with serious social problems that may have been kept locked away in a closet.

A key effect of the family disease of alcoholism was Bill W.'s orientation toward others. From a young age, he wished to help and save the world, and he began to love other people to a greater degree than he loved himself.[582] Such characteristics are found in people who have grown up in alcoholic environments. Before his grandparents took over, Bill W. was raised in a drinking milieu, long before he ever picked up a drink.

Only later, after his spiritual awakening, did Bill W. learn that in order to change others, and the world, we must begin by changing ourselves. His discussion of Step Eight says in part that, while making amends to others is very important, we must not forget to complete Step Four first, to learn all that we can about our own basic flaws and faulty relations with others that were often the ultimate source of our problems.[583]

These are masterful insights by Bill W., with no training in the helping professions but a great deal in the school of hard knocks. It is important to keep his words alive in the pages of Twelve Step history and in the practice of Twelve Step Higher Powered Mutual Aid principles in all our affairs.

---

581 *'PASS IT ON': The story of Bill Wilson*, 13–37; *Bill W.: My First 40 Years*, 1–17.
582 *Alcoholics Anonymous Comes of Age*, 107; *'PASS IT ON': The story of Bill Wilson*, 265; Al-Anon Family Groups, *First Steps: Al-Anon . . . 35 Years of Beginnings*, 156.
583 *Twelve Steps and Twelve Traditions*, 80.

Alcoholics Anonymous Comes of Age (1957)

Alcoholics Anonymous Comes of Age: A Brief History of A.A., Bill W.'s next major work, expounds upon the relationship between individual and social change. Again, he uses his own experience, taking us ". . . back to the time when I was a child, the time when I acquired some of the traits that had a lot to do with my insatiable craving for alcohol."[584] Bill W. knew, from his own life and that of others, that children can be strongly affected by the family disease of alcoholism.

Further along, he delves into more detail about the subject of one's upbringing, recognizing the effects on children of "faulty parental relations," while being careful not to blame.[585] At the same time, he describes the intergenerational interweaving of the family disease of alcoholism.

Clearly differentiating between individual recovery in A.A. and individual responsibility as regards outside issues, Bill W. says first that recovered A.A. members, "now restored as citizens of the world," would not "back away from their individual responsibilities" outside of A.A., in the world around them.[586] Secondly, Bill W. clearly reminds us that, "*As A.A.'s,*" members cannot take sides on any issue outside the purposes of A.A.[587] If that were to happen, it might well be the beginning of the end for A.A., as it was for other organizations broken apart by taking sides on external issues.[588]

Alcoholics Anonymous Comes of Age describes the difference between the single purpose inside of Alcoholics Anonymous, and responsibilities outside of it; here Bill W. states unambiguously that:

---

[584] *Alcoholics Anonymous Comes of Age*, 52.

[585] Ibid, 230.

[586] Ibid, 124.

[587] Ibid

[588] The Washingtonian Society, for example, was ruined largely by its involvement in outside issues. (*Alcoholics Anonymous Comes of Age*, 77, 124-25).

On Evolution

Our society will prudently cleave to its single purpose: the carrying of the message to the alcoholic who still suffers. Let us resist the proud assumption that since God has enabled us to do well in one area we are destined to be a channel of saving grace for everybody. On the other hand, let us never be a closed corporation; let us never deny our experience for whatever it may be worth to the world around us. Let our individual members heed the call to every field of human endeavor. Let them carry the experience and spirit of A.A. into all these affairs, for whatever good they may accomplish.[589]

Bill W. proved the courage of his convictions, attempting to enlist for the Second World War in 1942.[590] He had already served in the World War I and, according to Cheever, was "bitterly disappointed" when he was turned down.[591] The global common welfare must come first, Bill W. well knew, and that without a large measure of peace and serenity in the world, not even A.A. could continue to exist.

OTHER WRITINGS BY BILL W.

Bill W. took the last part of the Twelfth Step very seriously in other works. While writing the major books of A.A., the prolific author also wrote more than 150 articles for A.A.'s monthly magazine, the *A.A. Grapevine*, many later compiled in *The Language of the Heart* (1988). It devotes a separate segment of Bill W.'s writings to the subject of "in all our affairs,"[592] chiefly about individual responsibilities outside of A.A.

His article, "Your Third Legacy,"[593] talks of the importance of the General Service Office and the *A.A. Grapevine* for carrying the message worldwide. Bill W. wanted to use the latter to publish a series of articles

---

589 *Alcoholics Anonymous Comes of Age*, 232-33.
590 *'PASS IT ON': The story of Bill Wilson*, 272.
591 Susan Cheever, *My Name Is Bill*, 169.
592 *The Language of the Heart*, 233–86.
593 Ibid, 126–28.

that he was writing, titled "Practicing These Principles in All Our Affairs."⁵⁹⁴ In fact, he thought that he might expand these pieces into another book that would deal with the issues of living in the world outside of A.A.⁵⁹⁵ According to the A.A. Archives, he later abandoned this idea and put some of the material into the book, *As Bill Sees It: The A.A. Way of Life* (1967).⁵⁹⁶

Practicing the principles in all our affairs is discussed in many places in *The Language of the Heart*. In the March 1947 article, "Dangers in Linking AA to Other Projects," Bill W. refers to A.A. members as "citizens of the world."⁵⁹⁷ The March 1960 article, "After Twenty-Five Years," the July 1960 piece, "AA Tomorrow," and the February 1961 article, "The Shape of Things to Come," are other examples. In the latter piece, Bill W. said that we can always "step up" the practice of the principles in the world around us.⁵⁹⁸ And in the June 1961 article, "Humility for Today," he cites excuses we sometimes use to prevent us from doing our best, including the eerie and divisive idea that, since we have God's guidance, God is only on our side.⁵⁹⁹

Contemplating his retirement in the November 1961 piece, "Again at the Crossroads," Bill W. wrote that, like every A.A. member, he had the explicit responsibility to become a "citizen of the world," to practice the principles of the program in his public life, beyond A.A.⁶⁰⁰ Backing this up, he wrote that he was exploring opportunities outside of A.A., somewhere that he might be able to make a helpful and meaningful contribution.⁶⁰¹

---

594 Ibid, 327.
595 Ibid, 327; Nell Wing, *Grateful to Have Been There*, 1992), 28.
596 E-mail from the *A.A. Grapevine*, February 24, 2011.
597 *The Language of the Heart*, 44.
598 Ibid, 322.
599 Ibid, 255.
600 Ibid, 327.
601 Ibid

On Evolution

The email from the *A.A. Grapevine* said that Bill W. ". . . became involved in a new and innovative idea for converting oil into electricity, which he was really excited about."[602] The email also advised reading Chapter 24 of *'PASS IT ON'* for a "good overview of Bill's business activities in the late 1950s and 1960s."[603] It also said that Bill W. returned to Wall Street, too, which he thought was not only good for A.A., but was "excellent therapy" for himself.[604]

In *As Bill Sees It*, Bill W. declared that A.A. members must do more than carry the message to other alcoholics.[605] He suggested that, after a time, practicing the principles of the Twelve Step program extends to the world, where A.A. members are expected to work on responsibilities as full citizens outside of A.A.[606] Bill W. stressed the importance of practicing the principles in all aspects of our lives.[607]

The American Public Health Association's Lasker Award, offered to Bill W. but humbly declined, was presented to Alcoholics Anonymous in 1951. Its prophetic citation was partly quoted by Bill W. in *Alcoholics Anonymous Comes of Age*. The excerpt predicted that Alcoholics Anonymous could be recognized as:

> [A] great venture in social pioneering which forged a new instrument for social action; a new therapy based on the kinship of common suffering; one having a vast potential for the myriad other ills of mankind."[608]

These words accurately foresaw that A.A. would become a model that reached far beyond the alcoholism for which it was first intended. As earlier

---

602 Email from the *A.A. Grapevine,* June 4, 2011.
603 Ibid
604 Ibid
605 *As Bill Sees It: The A.A. Way of Life*, (New York: A.A. World Services, Inc.), 21.
606 Ibid, 21, 145, 262.
607 Ibid, 94, 123, 190, 265.
608 *Alcoholics Anonymous Comes of Age*, 301.

noted, Makela observed that, as of 1991, 260 different organizations were using the Twelve Step model,[609] originally constructed by and for Alcoholics Anonymous. White enumerated more than 400 such bodies in 1998.[610] Further progress on this front awaits the efforts of those afflicted by the many other profound ailments that now affect all of humanity.

It is of course the Twelve Step model, rather than A.A. itself, that can be used to solve "the myriad other ills of mankind." While Keating feels that "...AA has to accept its mission as a spirituality, not only for alcoholics, but for our time, since it addresses, head-on, the human condition across the board . . .,"[611] A.A. must be careful to restrict itself to carrying the message to alcoholics. Bill W. was very clear about this in *Alcoholics Anonymous Comes of Age*[612] and elsewhere, for example, in his February 1958 article, "Problems Other Than Alcohol."[613]

Alcoholics can suffer from additional disorders, too, as do many others. Hornbacher notes that psychiatry lists addiction as a mental disorder,[614] and ties mental illness and addictions together in her *Sane: Mental Illness, Addiction, and the Twelve Steps*. "All who were close to Bill W.," Robertson says, "were familiar with his swings from almost manic elation to the gloomiest depths."[615] A.A. worker Nell Wing said that he "described his depressive episodes as manic-depressive,"[616] currently named bipolar disorder. Bill W.'s depressions are described in A.A.'s biography, 'PASS IT ON'.[617]

---

609 Klaus Makela et al, *Alcoholics Anonymous as a Mutual-Help Movement*, 1996, 216.
610 William White, *Slaying the Dragon*, 1998, 163.
611 Thomas Keating, *Divine Therapy and Addiction*, 170.
612 *Alcoholics Anonymous Comes of Age*, 232-33.
613 *The Language of the Heart*, 222-25.
614 Marya Hornbacher, *Sane: Mental Illness, Addictions, and the 12 Steps*, 3.
615 Nan Robertson, *Getting Better: Inside Alcoholics Anonymous*, 39.
616 Nell Wing, *Grateful to Have Been There*, 1992, 53.
617 *'PASS IT ON': The story of Bill Wilson*, 36-37, 292-303.

On Evolution

However, Bill W. might also have suffered from post-traumatic stress disorder (PTSD). "Birth trauma" is sometimes cited as contributing to difficulties later in life;[618] Bill W. almost died at birth.[619] The birth trauma suffered by both him and his mother,[620] may well have led to a lack of mothering and his severe ups and downs. And he was nearly killed by accidental artillery fire in the First World War, saying that it was his "first experience with death close at hand."[621]

Bill W.'s emotional life was sometimes dominated by "power-driving," what he felt was his greatest character defect, synonymous with grandiosity and egotism. He learned, "starting with…revelations…about himself, that alcoholic drinking may…mask deeper psychological and emotional disturbances."[622] Bill W. saw therapists for a time,[623] making his stellar performance as A.A.'s cofounder even greater.

His partial autobiography, *Bill W.: My First 40 Years* (2000), the official A.A. biography of Bill W., *'PASS IT ON,'* and other books give accounts of Bill W.'s thinking on the subject of problems directly related to alcoholism. These works show, again, that he was well aware of the influences that contributed to his own drinking.

Echoing his words in *Alcoholics Anonymous Comes of Age*, in *Bill W.: My First 40 Years*, he began:

> …with childhood recollections bearing upon my background and ancestry and the events of that time as they related primarily to

---

618 See Otto Rank, *The Trauma of Birth*, (New York: Harper and Row, 1973; first edition 1929); Esther Menaker, *Otto Rank: A Rediscovered Legacy*, (New York: Columbia University Press, 1982); Bertil Jacobson, "Obstetric Care and Proneness of Offspring to Suicide as Adults: A Case-Control Study," *British Medical Journal* 317, no. 1346 (November 14, 1998).

619 *'PASS IT ON': The story of Bill Wilson*, 13. This is the opinion of an expert on birthing.

620 Expert opinion.

621 *Bill W.: My First 40 Years*, 54.

622 *'PASS IT ON': The story of Bill Wilson*, 299.

623 Ibid, 295–96, 334–35. Professional "outside help" has always complemented Twelve Step programs. See *If you are a professional…Alcoholics Anonymous wants to work with you*, pamphlet, (New York: A.A. World Services, Inc. 1986).

my personality structure and the defects in it, which no doubt laid the groundwork for my alcoholism.[624]

As the premier Twelve Step architect and author, Bill W. helped out in the writing of Al-Anon's Twelve Traditions[625] and Al-Anon's first book, *The Al-Anon Family Groups*.[626] Surely his contribution to the Al-Anon book included the statement: "Very often basic personality flaws can be traced back to the alcoholic's childhood."[627]

Bill W.'s paternal grandfather was "a very serious case of alcoholism,"[628] and his father, while perhaps not an alcoholic, was "at times a pretty heavy drinker."[629] It is evident that, as a child, Bill W. was affected by the disease of alcoholism well before he started drinking himself. He may thus have qualified for Al-Anon membership, too; alcoholics often grow up in homes where there is heavy drinking. In March 1958, Bill W. wrote that the children of alcoholics can develop emotional problems.[630]

These writings and more showed that Bill W. and others were well aware of the problems that occur in alcoholic homes. It was a small step for him to see that restoration to full citizenship by A.A. members included carrying out responsibilities, not only for A.A., but also for families, communities, countries, and the Earth. Says Marya Hornbacher in her book, *Sane*, after a time in A.A. she had to shoulder outside responsibilities, something to contribute to the world around her.[631]

---

[624] *Bill W.: My First 40 Years*, ix.
[625] *Lois Remembers*, (Virginia Beach, VA: AFG Headquarters, Inc.), 177.
[626] Ibid, 180.
[627] *The Al-Anon Family Groups, Classic Edition*, (Virginia Beach, VA: AFG Headquarters, Inc.), 20.
[628] *Bill W.: My First 40 Years*, 6.
[629] Ibid, 10.
[630] *The Language of the Heart*, 187.
[631] Marya Hornbacher, *Sane*, 85-86.

Al-Anon Twelfth Step work "in all our affairs"

It is no accident that alcoholism is called a family disease, affecting the relatives and friends of alcoholics, too. A.A. recognized this directly in its pamphlet "Is there an Alcoholic in your Life? AA's message of Hope."[632] Al-Anon points out that one doesn't have to drink to endure alcoholism.[633] Alcoholism might even be called a community, national, or international disease, such is its geographic and historic reach.[634]

Bill W.'s wife, Lois W., the cofounder of Al-Anon, stated her own beliefs in the world-changing potential of Twelve Step programs. She said, "It could really happen by people living [with] these principles,"[635] and "AA, Al-Anon, and Alateen [are] a big spiritual movement . . . a big factor in changing the world."[636] The Responsibility Pledge in Al-Anon is one example of these principles, and the Twelfth Step is another.

Al-Anon calls the Responsibility Pledge the "Al-Anon Declaration"; it is similar to A.A.'s Responsibility Declaration, and reads:

Let It Begin with Me.
When anyone, anywhere, reaches out for help,
let the hand of Al-Anon and Alateen
always be there, and - *Let It Begin with Me.*

Let It Begin with Me places responsibility for carrying the message with each Al-Anon member. "When anyone, anywhere, reaches out for help"

---

632 Al-Anon is mentioned on page 11.

633 Al-Anon Family Groups, "Are You Concerned about Someone's Drinking?" leaflet, (Virginia Beach, VA: AFG Headquarters, Inc.), 1981.

634 Browne-Miller, as noted, says that addictions have become a critical global problem (Angela Browne-Miller, xxv–xxvi).

635 *First Steps: Al-Anon…35 Years of Beginnings* (Virginia Beach, VA: AFG Headquarters, Inc., 1986), 156.

636 *Lois W. and the Pioneers*, two audiotapes, (Virginia Beach, VA: AFG Headquarters, Inc., 1982).

means just what it says. While Al-Anon gives priority to those affected by another's drinking, boiled down to its essence, the Al-Anon Declaration is also about helping all others. In practicing the principles "in all our affairs," Al-Anon members may simply use their new way of life as a model for those seeking sobriety, sanity, and serenity.

Al-Anon's Twelve Steps differ from A.A.'s by only one word: in its Twelfth Step the word "others" replaces A.A.'s "alcoholics." Al-Anon's Twelfth Step is quoted here partly because it is more inclusive than that of A.A., and partly because the greater part of my experience has been in Al-Anon; it reads:

> Having had a spiritual awakening as the result of these steps, we tried to carry this message to others, and to practice these principles in all our affairs.

The full import of Al-Anon's Twelfth Step was evident to me early on. I concluded that it was here that use of the Twelve Step Higher Powered Mutual Aid model had so much further potential to change ourselves and change the world.

Yet as I became more familiar with the program, it came as a surprise that the Al-Anon literature contained few references to the last part of the Twelfth Step, the practice of Twelve Step principles in all aspects of our lives. Further, few speakers or discussion meetings in Al-Anon focused on the phrase "in all our affairs," and how individuals were fully practicing a new, moral way of life, truly "living" the entire Twelve Step Higher Powered Mutual Aid program. That apparent failure is partly what drew me more deeply into the earlier literature of Alcoholics Anonymous.

AL-ANON LITERATURE

The powerful approach by the senior program, A.A., practicing the principles "in all our affairs," was paraphrased earlier from the 1953 *Twelve Steps and Twelve Traditions*. The 1981 *Al-Anon's Twelve Steps and Twelve Traditions*, published almost thirty years later, speaks of "following Al-Anon

On Evolution

principles each day and each hour . . ."[637] The 2005 revised version drops "each hour" from the text.[638] It speaks, too, of using Twelve Step principles "in all our affairs"[639] and as a "way of life,"[640] but without expanding.[641] Perhaps a decision was made to make Al-Anon's interpretation of "in all our affairs" more general than that of A.A. However, one result was that Al-Anon's explanation of the subject was not nearly as strong, and, in this respect, Al-Anon seemed like "A.A.-lite."

Al-Anon's daily readers, three in all as of 2002,[642] are a mainstay for reading and discussion meeting subjects. The index to the first, *One Day at a Time in Al-Anon*, has six references to the Twelfth Step, one of which says that a better way of living can be found in Al-Anon.[643] A second page addresses the question of how to apply the program in every aspect of living.[644] And a third asks the reader what can be done to improve her/his life, and restore herself/himself to "full citizenship" in global matters.[645]

The index to the second daily reader, *Courage to Change*, has only four references to the Twelfth Step,[646] none of which address in all our affairs.

---

637 Al-Anon Family Groups, *Al-Anon's Twelve Steps and Twelve Traditions* (Virginia Beach, VA: AFG Headquarters, Inc., 1981), 78.

638 *Al-Anon's Twelve Steps and Twelve Traditions*, revised 2005, (Virginia Beach, VA: AFG Headquarters, Inc.), 76.

639 Ibid, 77-78.

640 Ibid, x, 9, 61, 87.

641 The book's index has six references devoted to carrying the message in Step Twelve, but none to practicing the principles in all our affairs.

642 Alateen, a part of Al-Anon, also has two daily readers: *ALATEEN-a day at a time* (Virginia Beach, VA: AFG Headquarters, Inc., 1992) and; *Living TODAY in Alateen* (Virginia Beach, VA: AFG Headquarters, Inc., 2001). The former has two pages re: "in all our affairs" (29, 261), the latter one page. (294).

643 Al-Anon Family Groups, *One Day at a Time in Al-Anon*, (Virginia Beach, VA: AFG Headquarters Inc., 2000), 335.

644 Ibid, 343.

645 Ibid, 133.

646 Al-Anon Family Groups, *Courage to Change*, (Virginia Beach, VA: AFG Headquarters, Inc., 1992).

James G. Duncan

The index to the third, *Hope for Today*, also has four pages referring to the Twelfth Step.[647] Only one of these *Hope for Today* entries address "in all our affairs" in a manner consistent with the passion for a better world demonstrated earlier by A.A.

Often overlooked is Al-Anon's robin's egg blue book, *As We Understood . . .* (1985), a spiritual and inspirational introduction to Al-Anon's Twelve Step Higher Powered Mutual Aid, offering some good ideas about using the Twelfth Step to practice the program in every aspect of living; one excerpt is this:

> For many of us, the true value of the program becomes obvious as we discover that we are able to practice these principles not only with our family members, but with our employers and co-workers, neighbors, and even casual acquaintances . . . , The Twelve Steps have allowed us to see . . . that God's will for us is to make a contribution to the peace of other people who are travelling this path with us.[648]

The little book had a great impact on me. Halfway through reading it, I decided I was "in," later able to say, "I found God in Al-Anon." This was my spiritual awakening.

The 1990 booklet, *A Pebble in the Pond: The Twelfth Step in Action*, is about public information, co-operating with the professional community, and carrying out institutional work. While these are carrying the message, there is little mention of "in all our affairs." Also appearing in 1990 was *In All Our Affairs: Making Crises Work for You*. Its pages contain much about dealing with personal crises such as domestic violence, but few address the "in all our affairs" duties of Al-Anon members, at home, at work,

---

647 Al-Anon Family Groups, *Hope for Today*, (Virginia Beach, VA: AFG Headquarters, Inc., 2002).

648 Al-Anon Family Groups, *As We Understood...*, (Virginia Beach, VA: AFG Headquarters, Inc., 1992), 230-31.

On Evolution

in politics, in religion, in science, in peace and justice activities, or in environmental matters.

Living With Sobriety: Another Beginning (1979), an Al-Anon booklet, spoke of how the program has helped many to discover a new and better way of living.[649] Al-Anon also published a booklet titled, *Homeward Bound* (1993), stories from members about how they were coping with interactions between the alcoholism of loved ones and treatment centres. Poignantly, one member shared how he and his spouse had to "fall in love again."[650] Another writer said that, after his wife's treatment, he still went to Al-Anon meetings to find those who are "living the program."[651] A third revealed that she and her two alcoholic sons were now practicing the program principles in all facets of their lives.[652]

The Al-Anon booklet, *When I Got Busy, I Got Better* (1994), delves into how practicing the principles helped relieve one member's personal difficulties, outside Al-Anon.[653] One member felt, initially, that she had only to *try* to carry the message and practice the program in all her affairs.[654] Another said that Al-Anon has helped her to go beyond "talking the talk" in her life to "walking the walk" in all her affairs.[655] While these three latter booklets do not go into detail about the harm done when living with alcoholism, such as child molestation, suicide, and physical abuse, they do raise awareness.

A subsequent Al-Anon book, *From Survival to Recovery* (1994), has a number of good interpretations of the Twelfth Step. The book mentions

---

649 Al-Anon Family Groups, *Living With Sobriety: Another Beginning*, (Virginia Beach, VA: AFG Headquarters, Inc., 1979), 6, 22-23, 25, 27, 33.
650 Al-Anon Family Groups, *Homeward Bound*, (Virginia Beach, VA: AFG Headquarters, Inc., 1993), 7.
651 Ibid, 12.
652 Ibid, 17.
653 Al-Anon Family Groups, *When I Got Busy, I Got Better*, (Virginia Beach, VA: AFG Headquarters, Inc., 1994), 12-13.
654 Ibid, 22.
655 Ibid, 31.

"in all our affairs" on a page titled "Tapping Other Resources,"[656] and on page 258. Al-Anon is seen by many to be a new way of living.[657] A breath of fresh air for one member was the realization that her practice of the principles in all her affairs truly meant *all*.[658] The book notes that the Al-Anon principles for living can apply to outside groups as well.[659] One Al-Anon member recounted how her Al-Anon service work gave her firm principles to live by, and an appetite for further participation. It carried over to other aspects of her busy life.[660]

Showing promise, the 1995 book, *How Al-Anon Works*, shares that when looking for a sponsor, members wanted someone trying to use the principles in their entire lives.[661] It speaks of practicing the principles in every aspect of our lives, guiding us in relations with co-workers, in legal disputes, in establishing rules at home, or in leading a business meeting.[662] This book's index has six entries re: "carrying the message," but none for practicing the principles in all departments of life.

A sequel to Al-Anon's *As We Understood...* (1985), *Having Had a Spiritual Awakening. . .* (1998), shows how Al-Anon members are striving to practice Al-Anon principles in all aspects of life. One woman found that the principles worked for her in being honest outside of meetings, too.[663] Others practiced the principles in their daily lives, one concluding that they produce wonders for members around the globe.[664]

---

656 Al-Anon Family Groups, *From Survival to Recovery: Growing Up in an Alcoholic Home*, (Virginia Beach, VA: AFG Headquarters, Inc., 1994), 10.
657 Ibid, 107, 151, 165.
658 Ibid, 232.
659 Ibid, 242.
660 Ibid, 249.
661 Al-Anon Family Groups, *How Al-Anon Works for Families and Friends of Alcoholics*, (Virginia Beach, VA: AFG Headquarters, Inc., 1995), 37.
662 Ibid, 64.
663 *Having Had a Spiritual Awakening . . .* , 70-72.
664 Ibid, 172.

On Evolution

Al-Anon's *Paths to Recovery* (1997) brought a welcome return to the pioneers' explicit, specific approach to "in all our affairs." It is widely used in Step Study groups. The discussion of Step Twelve says that practice of the principles in all facets of our lives includes within families, with friends, in employment, and in community organizations, all important in our recovery programs.[665]

The 1997 book has thought-provoking questions about working each Step and Tradition, such as "What would change if I viewed service as a goal in every area of my life?" and "In what areas of my life do I need to start practicing these principles?"[666] The index to the book has many further references to the Twelfth Step and to Twelfth Step work, but these are mostly about carrying the message, not working the principles. (Al-Anon published a companion booklet titled, *Paths to Recovery Workbook* (2017), offering additional space to answer the questions posed in the book, *Paths to Recovery*.)

Paths to Recovery provides details about practicing Twelve Step principles in all areas of living, in bereavement, poverty, religion, and at work.[667] Some of these important words and ideas may be taken from A.A.'s questions published in the 1953 *Twelve Steps and Twelve Traditions*, paraphrased below for comparison purposes:

- Might we carry the principles of the program into our daily occupations?
- Might we bring new purpose and devotion to the religion of our choice?
- Might we accept poverty, sickness, loneliness, and bereavement with courage and serenity?

---

665 Al-Anon Family Groups, *Paths to Recovery: Al-Anon's Steps, Traditions and Concepts* (Virginia Beach, VA: AFG Headquarters Inc., 1997), 122.
666 Ibid, 128.
667 Ibid, 121-23.

However, *Paths to Recovery* falls short, and does not include a word or a paraphrase of one additional important question from A.A.'s *Twelve Steps and Twelve Traditions*:

- Might we meet our responsibilities to the world as well?

The Al-Anon Family Groups: Classic Edition (2000) addresses the subject of in all our affairs in at least four places. From "practicing all Twelve Steps in daily living,"[668] to "all departments of daily living,"[669] the book concludes, "It is a lifetime job to apply the Steps to our everyday lives."[670] On the page referring to Step Twelve, however, there is no mention of "in all our affairs."[671] In the book's Appendix II, historical information about Step Twelve includes practice of the principles "in all departments of daily living."[672] The 1966 edition of the same book says that recovery "...brings with it increased capacity for honesty, unselfishness, and love in all areas of daily living."[673]

Al-Anon's 2007 book, *Opening Our Hearts, Transforming our Losses*, is partly about using the program in other areas besides alcoholism.[674] It mentions "in all our affairs" in at least two places,[675] without expanding, but notes how we can be a living model to others by putting the Al-Anon

---

668 *The Al-Anon Family Groups: Classic Edition*, (Virginia Beach, VA: AFG Headquarters, Inc.), 36.

669 Ibid, 119.

670 Ibid, 120.

671 Ibid, 61.

672 Ibid, 119-121.

673 Al-Anon Family Groups, *Living with an Alcoholic* (Virginia Beach, VA: AFG Headquarters Inc., 1966). 52. The book's title had been changed to *Living with an Alcoholic* in 1960, but it was re-instated as *Al-Anon Family Groups* in 1984. (*The Al-Anon Family Groups: Classic Edition*, 2000, 4).

674 Al-Anon Family Groups, *Opening Our Hearts, Transforming Our Losses*, (Virginia Beach, VA: AFG Headquarters Inc.), 90, 152.

675 Ibid, 103, 109.

principles into practice.[676] Promisingly, the book says further that the tools we are given in Al-Anon help us to lead more fruitful lives, at work, with family and friends, and in our romantic relationships.[677] Dealing directly with Step Twelve, one member said that he is creating a better life for himself, partly by helping others in the same boat.[678]

In *Discovering Choices*, a 2008 Al-Anon book, Step Twelve is mentioned in regard to service work, carrying the message.[679] The volume also speaks of integration of the principles in our everyday lives[680] and improvement in outside relationships of all kinds,[681] including with families, at work, and in all other aspects of life.[682]

Many Voices, One Journey (2011), partly an account of Al-Anon history, has a number of references to the interpretation of the latter part of the Twelfth Step.[683] This includes a comment from Lois W., that Twelve Step programs are an unquestionable force for good in global affairs.[684] This consolidated her opinion, noted earlier, that "AA, Al-Anon, and Alateen [are] a big spiritual movement . . . a big factor in changing the world."[685]

Al-Anon's workbook, *Reaching for Personal Freedom* (2013), speaks in the preface of how the Twelve Steps pertain to other facets of our lives, beyond alcoholism.[686] It says as well that the wisdom of the Al-Anon

---

676 Ibid, 154.

677 Ibid, 152.

678 Ibid, 170.

679 Al-Anon Family Groups, *Discovering Choices: Recovery in Relationships*, (Virginia Beach, VA: AFG Headquarters Inc.), 243.

680 Ibid, 243.

681 Ibid, 245, 250, 258, 301–23.

682 Ibid, 303.

683 Al-Anon Family Groups, *Many Voices, One Journey*, (Virginia Beach, VA: AFG Headquarters Inc.), 3, 22, 88.

684 Ibid, 27.

685 *Lois W. and the Pioneers,* two audiotapes, (Virginia Beach, VA: AFG Headquarters Inc., 1982).

686 Al-Anon Family Groups, *Reaching for Personal Freedom: Living the Legacies*, (Virginia Beach, VA: AFG Headquarters Inc., 2013), 7.

program can give us growth in understanding, specifically regarding the practice of the principles in all our affairs.[687] Included in the discussion of the Twelfth Step is an *apropos* sentence, which says that our practice of the principles in all aspects of life helps us to embark on a journey of personal transformation in recovery and healing.[688]

For one Al-Anon member, the principles had become the key for living that she practices in every facet of her life.[689] The Al-Anon program, said another, has transformed his life: relationships of all kinds have gotten better, difficulties are settled more easily, and he is more accepting.[690] Wrote another, though the biggest problem in Step Twelve is practicing the principles in all aspects of her life, it has become, with time and effort, her permit to a better life.[691]

Interspersed in the 2013 workbook are examples of the principles that might be practiced "in all our affairs." For instance, in the section on Steps Two and Three, we are informed about the difficulties and rewards of having a Higher Power in our lives, inside and outside of meetings.[692] In the discussion of the Twelfth Tradition, one contributor shared how she used the Al-Anon principle of anonymity in her work as a journalist.[693] Keeping the principles in mind, a different writer explained her relationship with money.[694] She said, too, that her life must be based on firm Al-Anon principles.[695]

The cumulative effect of having this large body of Al-Anon Conference Approved literature, addressing the latter part of Al-Anon's Twelfth Step, is that there is a reasonably complete description of the practice of the

---

687 Ibid
688 Ibid, 61.
689 Ibid
690 Ibid, 63.
691 Ibid, 65.
692 Ibid, 15-24
693 Ibid, 115
694 Ibid, 169.
695 Ibid

On Evolution

principles "in all our affairs." The Al-Anon literature has many references to this part of the Step, most of a general nature, not weightier, specific issues such as child or spousal abuse, sexual assault, suicide, or gun play that can sometimes, or often, be a part of living with alcoholism.

There is a need to address other particular outside issues, such as climate change, ethnic cleansing, human rights, overpopulation, pollution, poverty, and racism - outside of Al-Anon's immediate issue of dealing with others' drinking. Al-Anon members, not Al-Anon, must take their fair share of responsibility for outside local, community, and world affairs, in all of God's Creation. Some of these specific issues are taken from a list drawn up by John Cavanagh in 2004.[696] If the current state of affairs brings an end to civilization, Twelve Step programs would perish as well. Bill W. knew this in 1942 when he tried, to no avail, to report for duty in World War II.

Mander writes up a list similar to Cavanaugh's,[697] saying that Indigenous peoples need much help, especially from established movements re: women's issues, the environment, anti-globalization, global justice, human rights, and democracy, etc., and "also from individuals becoming active on a broader scale."[698] However, neither Cavanagh nor Mander refer to Twelve Step programs that can transform ourselves and transform the world. They fail, too, to tap into Indigenous peoples' healing methods, and there is no mention of spirituality that would be of much assistance.

Future Al-Anon books and editions can be more specific about practicing the principles on God's still-green earth "in all our affairs." For ease of reading about and understanding the Twelve Step way of life, it would be useful for Al-Anon to publish a pamphlet or book that might

---

696 John Cavanagh et al, eds., *Alternatives to Economic Globalization: A Better World is Possible*, (San Francisco: Berrett-Koehler Publishers, 2004), 333-45.

697 Jerry Mander et al, eds., *Paradigm Wars: Indigenous Peoples' Resistance to Globalization*, (San Francisco: Sierra Club Books), 2006.

698 Ibid, 226.

be called, *Al-Anon's "In All Our Affairs."* The same might apply to A.A., the perennial leader of the Twelve Step movement.

"In all our affairs": Confusion and Uncertainty?

Noted earlier, 7 of the 12 Steps deal with morality. Step Four asks us to make a "searching and fearless moral inventory of ourselves." Step 5 says we need to admit "the exact nature of our wrongs." In Step Six we are to be "entirely ready to have God remove all these defects of character." In Step Seven, we humbly ask God "to remove our shortcomings." Step Eight asks us to make a list of those to whom we owe "amends," and become willing to make those amends. In Step Nine we are to make "direct amends." if possible, and Step Ten cautions us to continue "to take personal inventory and when we were wrong promptly admitted it."[699] At first then, there is little room for listing our qualities and talents, not as necessary as acknowledging our defects of character, flaws, and shortcomings. In the end, some balance is often achieved.

We can adopt Abrams's "planetary morality"[700] in HPMA programs that separate good from bad, by practicing the Twelve Step principles in all aspects of life. An important action to take here is "voluntary simplicity,"[701] to enable many to end immoral over-consumption of all kinds, especially in developed countries who have large ecological footprints out of all proportion to their population size.

While A.A. and Al-Anon need to provide leadership to illustrate these considerations for Twelve Step devotees, it is up to members, individually and/or in groups, to make a "searching and fearless" examination of their efforts in working the entire Twelfth Step, "living the program" in all

---

699 See A.A.'s Twelve Steps at the front of the book.

700 Nancy Abrams, *A God That Could Be Real: Spirituality, Science, and the Future of Our Planet*, 2015.

701 First used by Richard Greg in *The Value of Voluntary Simplicity*, (Wallingford, PA: Pendle Hill, 1936).

On Evolution

aspects of life. This approach is not new, but one has to dig in Al-Anon to find the gems of "in all our affairs" in action.[702]

One alcoholic related that members' practice of the program in their daily lives, not just in talk at meetings, attracted him to A.A.[703] Said another, "A.A. is not something I joined; it's something I live."[704] Through such informed opinions, we can determine what issues should be included in the holistic phrase "in all our affairs." The catch, again, is that "all" does not leave any room to wriggle out of responsibility, none in fact.

Use of the word "all" implies adopting an all-encompassing, holistic, ecological approach to the practice of Twelve Step HPMA principles, from bottom to top and side to side, from "individuals in relation" to outer space, from Earth to the Almighty. Armed with this understanding, practitioners of Twelve Step HPMA - outside their first objectives of recovery and healing - can adopt an approach that is similar to the practice of ecological, healing, holistic First Nations Higher Powered Mutual Aid.

However, there seems to be among many a block to taking responsibility for the practice of the entire Twelfth Step, and since the passing of the pioneers, adequate leadership from A.A. and Al-Anon has been lacking. Closer study of the work of the founders on this subject, it is suggested, might be part of a new beginning for all Twelve Step members and their societies. The pioneers were clear about practicing the principles "in all our affairs," both inside and outside of meetings. The Twelve Step principles that we practice are no different anywhere, in groups or outside of groups.

Uncertainty can be created, though, if Twelve Step members understand that there is, or might be, a difference between principles practiced in meetings and principles practiced outside meetings. Confusion can arise when outside issues in meetings are seen as outside

---

702 See, for example, *Al-Anon's Twelve Steps and Twelve Traditions*, (Virginia Beach, VA: AFG Headquarters, Inc., 1981), ix-x; and Al-Anon's *Opening Our Hearts, Transforming Our Losses*, (Virginia Beach, VA: AFG Headquarters, Inc., 2007), 90, 152.

703 Alcoholics Anonymous, *Daily Reflections: A Book of Reflections by A.A. Members for A.A. Members*, (New York: A.A. World Services, Inc., 2009), 60.

704 Ibid, 63.

issues after meetings are over. Worse, if such confusion persists, one might develop a false principle that says Twelve Step members should not be concerned about outside issues at all! The Twelve Step pioneers would be very disappointed at any denial or misunderstanding over this critical matter. Working the entire Twelfth Step should not be a source of uncertainty or confusion.

The whole object of Twelve Step Higher Powered Mutual Aid, taking us beyond recovery and healing, is practicing healthy, holistic, positive principles of moral living for use in participation at meetings, and after we return to our families, communities, and the outside world. However, it is only possible to adopt these principles after we have learned to focus on and love ourselves. Concentrating on issues in our individual lives is the most we can do at first. But the wider implications of the last part of the Twelfth Step, cannot be shuffled off to the side or forgotten after the good ship has been righted.

A long standing "potent quietest element" in A.A., says Raphael,[705] may be militating against taking action in all our affairs. This is perhaps because A.A. and other Twelve Step societies can use only attraction, not promotion of their programs.[706] He quotes one writer to suggest a reason for putting some problematic aspects of our lives on a back burner - or off the stove altogether. In the case of alcoholics, quotes Raphael, "Surviving the passage through active alcoholism...renders humble acquiescence to everyday life contrastingly so superior to dissonance of any sort that wilful disruption is, for many, wholly unthinkable."[707] Dealing with this earlier, in 1964, Gelman said "...the alcoholic must give highest priority to...factors which help maintain his sobriety and must guard against those circumstances which may threaten his stability."[708]

---

705 Matthew Raphael, *Bill W. and Mr. Wilson*, 55.
706 See A.A.'s Tradition Eleven at the front of the book.
707 Matthew Raphael, *Bill W. and Mr. Wilson*, 55.
708 Irving Gelman, *The Sober Alcoholic: An Organizational Analysis of Alcoholics Anonymous*, (New Haven, CT: College and University Press, 1964), 171.

On Evolution

For the families and friends of alcoholics, surviving the family disease and finding a safe haven in Al-Anon makes engagement with the insanities of the outside world less inviting. The world can be a dangerous place, and deliberately involving oneself in contentious issues may place at risk hard-won sobriety, sanity and serenity. If at all possible, we use the principles of the program to let go of, or detach from, people and issues over which we appear to be powerless.

Yet, as Scott Peck said in *The Road Less Traveled*, life is a risk.[709] Choices that we make are always imperfect and therefore entail taking chances. Thus, it does not make sense to say that our powerlessness always brings inaction, even over contentious issues. If that were the case, why would we go to Twelve Step meetings in the first place; use the gifts of the program to make a better life for ourselves and others; pray and meditate; vote; say and do what we believe in; model our lives on program principles; and stand up to be counted in our families, communities, countries, and the world?

As Al-Anon's *Hope for Today* succinctly says, practicing the program "in all our affairs" can keep us coming back to meetings.[710] At meetings - and elsewhere - we must be able to talk about the results of applying the Twelve Step principles, living the program in every facet of our lives. Said one alcoholic, if he worked the Steps in all his affairs, he would have a life that he could only dream about.[711]

The promise is great: one professional, an A.A. member, said that there is still much "hidden potential" in the practice of Twelve Step principles in all our affairs.[712] Over 80 years since A.A.'s founding, and over 65 years for Al-Anon, their potential has not yet been attained, not at all. Noted earlier was Browne-Miller's 2009 contention that there are 2 *billion* alcohol users in the world, 1000 times greater than the current A.A. membership.

---

709 M. Scott Peck, *The Road Less Traveled*, (New York: Simon & Schuster, 1978), 134.

710 *Hope for Today*, (Virginia Beach, VA: AFG Headquarters Inc., 2002), 74.

711 *Daily Reflections*, (New York: A.A. World Services, Inc., 1990), 54.

712 "A Member's Eye View of Alcoholics Anonymous," pamphlet, (New York: A.A. World Services, Inc., 1970), 25-26.

What is to be done? Time has proven that Tradition Eleven is too restrictive.[713] With a change in the Tradition, Twelve Step practitioners would be able to *promote* their programs in the outside world, rather than use only attraction. It may be that the Twelve Step pioneers would agree that a change such as this - in view of the lack of overall success - can be made to ensure that Twelve Step programs are more accessible to all.[714] Again, a change such as this could begin with a single group conscience (or decision).

The notion that it is "wholly unthinkable" to involve oneself in outside issues appears weak and incorrect. Timing is everything. Early on in recovery, we may feel powerless over difficult issues. However, later, with sufficient sobriety, sanity, and serenity under our belts, including such issues in our daily lives, even through prayer and meditation, may be necessary for completion of the Twelfth Step. Prayer and meditation in Step Eleven practice can be used to help guide us along, to counter what may seem impossible for us to carry out in our daily lives.

Keating writes of the power of, and necessity for, prayer and meditation in A.A., to change ourselves and change the world. He says that the practice of Twelve Step principles "refers to a transformation of consciousness that penetrates and encompasses the whole of our lives and all our relationships: God, other people, ourselves, planet Earth, and the cosmos."[715]

Let there be no doubt that if it were up to me, in a dictatorial mood, the practice of holistic, Higher Powered Mutual Aid principles "in all our affairs" would be clearly defined in the humanly knowable truths of the individual in relation to the common welfare, leaving the remainder to the Higher Powers, the guardians of ultimate truth. *Alcoholics Anonymous*

---

713 See A.A.'s Twelve Traditions at the beginning of the book.

714 Bill W., as noted, hoped that pioneering would never end. (*Alcoholics Anonymous Comes of Age*, 80). Here, Bill W. wrote of the need to bring A.A. to more of the world. But I take it to mean, also, pioneering within structures such as A.A.'s Twelve Traditions.

715 Thomas Keating, *Divine Therapy and Addiction*, 157.

On Evolution

Comes of Age shows us the pioneers' passion for a better, healthier world, a passion that we also need to have; it says, again, that:

> [T]he members of Alcoholics Anonymous, now restored as citizens of the world, are [not] going to back away from their individual responsibilities to act as they see the right upon issues of our time.[716]

Bill W., truly a world citizen, used the same kind of language elsewhere, as noted, for example, when contemplating his retirement.[717] If in fact we are restored to sobriety, sanity, and precious serenity, we can "step up" the practice of Higher Powered Mutual Aid principles and programs in all areas of our lives.

Of course, neither A.A. nor Al-Anon can go about listing all the aspects of life in which the principles might be practiced; such a record would be endless. Practicing the principles "in all our affairs" is mainly an "individual in relation" practice in and outside of meetings, and less a practice by Twelve Step groups or societies themselves.

However, there is no reason why Twelve Step members, committed to practicing the principles in all facets of life, cannot join with other Twelve Step members in groups that want to deal with issues of broader family, community, national, or global concern.

To be based on a solid HPMA foundation, and to withstand the vagaries of the outside world, Twelve Step leadership would be needed in these "outside groups" that would ensure the practice of Twelve Step principles in the groups. Twelve Step members, forming outside groups, would consist of those well versed in Twelve Step philosophy and methods, particularly in the practice of the principles of morality, self-regulation, and self-responsibility. As important, we would need to make sure that the "outside groups" of Twelve Step members stick to their mandate, so that the issue is dealt with effectively.

---

716 *Alcoholics Anonymous Comes of Age*, 124.
717 *The Language of the Heart*, 327.

"Outside issues" of the Twelve Step "outside groups" should be avoided, to ensure the group's lasting prosperity. Avoiding issues "outside" of broader family, community, national, or global concern, is the same as avoiding "outside issues" at Twelve Step meetings. For example, groups of Twelve Step members dealing with climate change would need to sidestep issues of ethnic cleansing, terrorism, and world conflicts that would divert them from their group's purpose. People wanting to join outside groups of Twelve Step members may need newcomer meetings, and to listen and watch as newcomers do at First Nations and Twelve Step meetings.

The election of leaders in these outside groups should be avoided, so that they "are but trusted servants," observing the guidance of the group. Authoritarian hierarchical organization needs to be prevented. Here too, Twelve Steppers might participate in First Nations circles to discover how they deal with issues related to holistic healing and recovery. First Nations people may want to study Twelve Step methods as well.

Personal Recovery

Why is it so important for me to harp on the phrase "in all our affairs" in the Twelfth Step, inspired by the A.A. and Al-Anon founders' early words and actions? First, I see that, by the grace of God, the application of the principles of HPMA has greatly improved my life and those of many others, and I have come to believe that it has the potential to change the world. Here, I am only in full agreement with the pioneers who accomplished their dream, starting the Twelve Step movement to battle alcoholism first, and then many other ills. They were justified in thinking and feeling that their work might actually change the world for the better.

Second, I look around me after over a half-century of life and see that things are going from bad to worse in my community, my country, and in the entire world. I see my HPMA circles and groups being affected in negative ways, and search for ways to improve them. I don't know if the world is now apocalyptic or apoplectic, and think I am being a realist,

On Evolution

not a pessimist. No rose-coloured glasses for me. I will not dwell on the evidence that "normal" gets worse each day or year at a time, but, instead, will try to keep the focus on myself and expand on more lessons learned from my own HPMA recovery.

Twelve Step "oldtimers" like myself, inclined to accept more and more of the program, can decide to swallow the bad with the good, like medicine given with honey as a child. The male God that has stared at me from Twelve Step pages has become part of the medicine that tastes better with the honey of the program. Safe in Twelve Step meetings, literature, philosophy, and therapy, I have taken risky limbs to find that honey, in wider views of Higher Powers.[718]

With the maturing of my program, I can only applaud each person who arrives at her or his own understanding of a God who guides her/him. This is honey in the program. Feeling more and more of the freedom that the Twelve Steps offer, my own conception of a Higher Power has evolved, and I can now conclude that the God of my understanding is not entirely knowable. As a human being, I ask: Who am I to define God, save - using faith, hope, and prayer - for benevolence and love?

To my mind, the riskiest limb is embodied in the Twelfth Step. The Twelfth Step, the A.A. Responsibility Declaration, and the Al-Anon Declaration, are all about carrying the message to alcoholics and others, and practicing the principles in all our affairs. I will try to describe what Twelfth Step work means to me, after years of learning and relearning the message in A.A. and Al-Anon groups, and in First Nations circles.

Developing a new and more fruitful way of living can be a long, arduous, but rewarding task. Trying to "live" the whole Twelve Step program has helped me change my self-defeating learned behaviours to the degree that my whole approach to living has been transformed. As a fortunate "high bottom" alcoholic - one who has not reached the almost bottomless depths of alcoholism - I ventured into A.A. and stopped drinking.

---

718 To avoid the gender bias of "God" I pray to the "Great Spirit," taken from a First Nations prayer.

Further progress awaits further effort on my part, but I might characterize the present results as:

- personal "repowerment,"[719] a better knowing and powering of myself and my purpose(s) within the common welfare.
- an increase in sobriety, sanity, and serenity, as I carry the message to alcoholics/others and practice the HPMA principles in all my affairs.
- a dutiful increase in my self-regulation and self-responsibility, by defining a consistent, moral place for myself in the imperfect social and environmental fabric of my life.

These results came about in a step-like process. Prior to receiving professional help, and to my participation in Al-Anon and A.A. groups, and in First Nations circles, I did not really know myself in a fundamental sense. Thus, any action that I took did not grow from the ground of my being, my purpose(s) for living. Having come to know myself to a much greater degree, I can more properly take action, with guidance provided by HPMA practice, appropriate for my existence as an "individual in relation" ecological being.

Consistently avoiding issues of control, fear, and trust made healthy intimacy impossible for me. Yet no longer the bad apple I thought I was, I now have a core of positive purpose(s) with crisper boundaries, fitting together better with people who have achieved similar results. Good enough apples that we were becoming, we could carry the message to others and live the HPMA program in all aspects of life.

Simmered down, my program is about respect, in three parts. This is the message that I try to carry with me to others and to the world around, in

---

719 I use the term repowerment, "powering again," in place of the popular but ill-defined "empowerment," "giving power to." The latter assumes that one's power has somehow been lost. Yet, one's power is never lost until death; it always remains, flickering away in the unconscious. Fanning the embers of our individual power into greater life therefore involves a *repowerment*, not an *empowerment*. Simos uses similar imagery, "fanning the flames of [a person's] creative fire" without using the word empowerment. (Miriam Simos, *The Earth Path*, 2004, 125).

On Evolution

all my affairs: respect for myself, respect for the other, and respect for what keeps us God's children. There is an additional note on serenity in my life.

Respect for myself

Outwardly successful and confident, I had little respect for myself when I walked into my first Al-Anon meeting. I thought I was in the right place, but my feet were in the wrong shoes, or the shoes were someone else's. But I worked with the best of them until the program had a chance to work its near magic in my life, so that now I can walk the talk to the best of my ability. The family disease of alcoholism had taught me to run before I could walk, to seek all knowledge to protect myself,[720] and when, of necessity, I had to adopt a slower, patient approach, "living the program" was difficult.

Learning that I had to focus on and take care of myself first, instead of others, came as quite a surprise, and only gradually did it become a way of life.[721] When I did start to turn that habit around, I likened it to turning the HMS *Queen Mary*, and I have not yet turned it completely around. But I found my own shoes on the deck, and they're on the right feet most of the time. The principles of Higher Powered Mutual Aid can teach me until the end of my days, and for that I am grateful as well as responsible.

All of this took a long time, and I could not have recognized the problems on my own. It took a glancing blow from a spiritual two-by-four, then a referral to Al-Anon by my therapist for me to begin to come to my senses. I can look back and compare it to tying my shoelaces together. Hobbled and staying in one place, I could manage only baby steps. With the assistance of my "HP," I knew that those shoelaces *had* to be tied together, so that I could stay with the program until I got the message.

Sure, I untied them a couple of times, galloped off in wrong directions, and found myself "nowhere man," at familiar dead-end destinations. But I came back and learned the lessons better. Good work habits kicked in

---

720 Finished a University degree at the age of 30, I told a younger sister, sick as I was, that "I now realize I will never know everything."

721 This is not to deny that helping others, too, is essential when working HPMA.

when I needed them most, and I was rewarded, like a wayward bear, with more and more of the honey of the program.

As an increasingly *human* being, abandoning alternating superhuman and subhuman roles, as well as the treadmill of the "human doing," I determined that I would not be able to please everybody and that not everybody would like me. This had been partly an unconscious intention to "know all and please all," so that I could be completely safe, away from the hazardous world that I'd learned about as a child. Controlling the world in such ways was far beyond my human limits. I could not be God after all, but I found that I could be me. That is perhaps the greatest gift of my program.

Respect for the Other

What does "respect for the other" actually mean to me? It simply means respect for all of Creation, the animate and inanimate, that which is seen and not seen, that which is heard and not heard, that which is tasted and not tasted, that which is felt and not felt, that which is understood and not understood . . . A tall order? No. Once I started to trust my Higher Power, myself, and others, and let go of the fear and the terror, the anger and the rage, the hatred and the ego, it was a piece of cake. I like cake and I thank my Higher Power that there are more pieces left, more lessons to learn, because I need them to continue discovering my new way of life.

At the early tender time in the program, I thought I had respect for the other. But this respect was really a strong desire to control by pleasing, to control by knowing, and to change the world to make it safe for myself. Focusing on others (i.e., the drinker) was a habit that I first learned as a toddler, and from that time on I was full of conscious and unconscious fears, "living in my head" to avoid dangerous feelings sealed safely away. As well, I did not respect other species. I was a disrespectful hunter in my youth, mainly for "sport." That, too, has changed, in great measure thanks to HPMA. Now I do not hunt.

Respect for What Keeps Us God's Children

This is a bit of a New Age mouthful, but what does it really mean? Briefly, it is my personal belief that the "individual in relation" members of Creation

On Evolution

are all children of God,[722] and that there is an ecological state of being that keeps us together as sisters and brothers; this includes what we call the inanimate. To "Keep It Simple," I again use the maxim that everything is connected to everything else. This worldwide, ecological weblike state is evident in the practice of HPMA principles, in and outside of meetings.

So now I have come to this wonderful place of respect. Why respect, of self, the other, and for what keeps us together as God's children, rather than love? Why not rely on "Love thy neighbour as thyself." Wouldn't that do it? "Respect" may simply be a necessary stage that I am going through in my recovery. Yet there are some deeper reasons why I prefer this three part respect. Much of it is about keeping it simple and methodical, so that I can learn the spiritual and ecological principles slowly and well.

In all honesty, I am leery of love. Alan Paton referred to the problem here, in the end, a spiritual one. At one point in his novel, *Cry, The Beloved Country*, he said, "Let him not love the earth too deeply."[723] Love, like hate, its opposite, can be an enmeshing addiction. That might be why I like respect better than love, at least for the time being.

As earlier, we can go to definitions. A Webster's "respect" is "to consider worthy of high regard," while one chosen from the Oxford complements it: "avoid interfering with, harming, degrading, insulting, injuring, or interrupting." A major problem for me upon entering the program was considering myself with high regard. I did not respect myself.

This was manifested in many ways, sometimes with self-disrespect; focusing on and/or disrespecting others; irresponsibility; grandiosity; depression; and defensiveness. So I had to learn self-respect for the first time. I had to focus on myself more thoroughly, to develop a core of purpose(s) that I could respectfully call "me," by listening and talking in the safe atmosphere of HPMA circles/groups where "interfering with,

---

722 One First Nations Elder said, "If we are children of God, there must have been a woman involved somewhere." Personal communication.
723 Alan Paton, *Cry, The Beloved Country* (New York: Scribner, 2003, first published 1948), 111.

harming, degrading, insulting, injuring, or interrupting" were not among the discussion guidelines.

Before the Higher Powered Mutual Aid way of life took hold, I unconsciously rehearsed things in my head before speaking, so that what came out of my mouth might be acceptable to others. And while listening to others, I formulated counterargument, as if preparing for debate. These were habits I did not consciously change. The adjustments simply resulted from my participation in HPMA meetings and all that that implies. I noted earlier that I began to speak from my heart, not only my head.

At about the same time, I began to learn that my early desire to "please all and know all" was not really about respect for others, but about fear and control. My pleasing and knowing were simply designed to allay fear of the consequences of not pleasing and not knowing. Under the pleasing/knowing exterior lay lack of trust, fear, and, yes, hatred, like an iceberg, mainly in the unconscious. With the help of the program, I was able to build trust of myself and others, let go of destructive emotions, and relinquish control.

Through the recovery process, too, I began to learn about boundaries, like the peel of the apple that fit around the emerging core of my positive, creative potential and purpose(s). Today, I think that I am able to respect to a greater degree the boundaries of others, as well as my own. I learned about "differentness," that each of us has unique positive potential(s) and purpose(s), complete with possibly unique Higher Powers, our own understandings of God.

The formation of healthy boundaries is partly a result of practicing the principle that I described in the third respect: "Respect for what keeps us God's children." This "in relation" state is exemplified in the Twelve Traditions and in the discussion guidelines. Rooted in individual freedom and the common welfare, this state is defined by respect, thereby teaching respect for the "individual in relation" to Creation.

What is the result of all this respect for me? It sounds so serious, so boring, so dull, like the word serenity did when I first took up my membership in Al-Anon. Don't we have a little fun once in a while? Well, I will keep this one short, and say that I used to have a lot of "fun" that was

On Evolution

harmful to me and to others. Oddly, I used to believe that I had to drink to enjoy myself.

SERENITY

When I entered the program, serenity was a foreign concept, even an unwelcome one. Still tilting at windmills like Don Quixote, I felt that courage should be first in the Serenity Prayer and serenity dead last. But I began to learn the value of serenity, and why A.A. and Al-Anon began to use the Serenity Prayer so early in Twelve Step history.[724] Though I couldn't figure out why it was important, I could experience serenity and feel it in First Nations and Twelve Step gatherings, and it slowly came unbidden into my life.

Why is the gift of peace - serenity - so precious, so priceless? Its high value may be related to its short supply. Serenity is one of the greatest benefits of working my program. In practicing the program in all my affairs, I must maintain my sobriety, sanity, and serenity if I am to be an effective citizen of the world around me. If I am to attract more people to HPMA programs, and keep them coming back, I can do my best to exemplify the program in every way possible, and to be sober, sane, and serene wherever I am.

I do not carry the jewel of peace and serenity away to a secret hiding place within me. Instead, I try to practice it in my daily life and take it to and from meetings, to share with others. Like any of the other gifts that I receive in HPMA circles and groups, I must give my serenity away in order to keep it front and centre. Practiced in all my affairs, serenity is, hopefully, an integral part of the message that I carry to others. By 2015, it was gratifying to hear one Twelve Step member say that I was "more calm" than I'd been in the past, and by a person outside the program, "serene."

---

724 The Serenity Prayer was found by an A.A. member, Jack C., in an obituary column of the New York *Herald-Tribune* (*'PASS IT ON'*, 252). Its origins are somewhat obscure (*'PASS IT ON'*, 258, note 6).

So I have now come to a place where I can use my heart and head more or less in tandem. The link between them is being reforged in a new cradle, HPMA, allowing me to rejoin the circle. I have learned to live a quieter, simpler life, figuratively somewhere between my heart and my head, and, emotionally, away from the extreme, disruptive feelings of earlier days. "Living in my head" had not worked, and I was able to bring healthy emotion into my life. Equally of course, "living in my heart" ignores the contributions that my head can make. I lead a more balanced life.

HPMAGs

I can't shake the notion that I've heard or seen or felt or known all this before. Perhaps it's somewhere deep inside at the molecular level, or in a strand of DNA. It may be an echo, reverberating down a long hall, or along the silent path I have found and tread upon more confidently with others. Hansen is correct when he says, that Twelve Step programs may "...tap into a natural process that we all share."[725] That "natural process" is Higher Powered Mutual Aid, with its self-regulation-enhancing programs. Anarchic self-regulation is derived from one's innate decision making "power from within," not power from outside sources.

Joseph Nowinski asks if A.A. is a "real solution" for alcoholism,[726] and goes on to prove it. Nancy Abrams wonders why her Twelve Step program works,[727] and goes on to answer her question. The idea comes to the fore that serenity is much more important than I'd thought. No doubt it has a history as long as, or longer than, that of violence.

We have always been social and ecological beings - individuals in relation - and HPMA groups (HPMAGs) have existed from the beginning. The HPMA way of life, along with competition that is not habitually destructive, is a healthy way to live, to help sustain us on Earth as useful, valuable parts of Creation. HPMAGs help us to learn and use real,

---

725 William Hammond III, *12 Step Wisdom at Work*, 95.

726 Joseph Nowinski, *If You Work It, It Works!* 2015, 5.

727 Nancy Abrams, *A God That Could Be Real*, xxvi.

On Evolution

spiritual, and holistic mutual aid, but such groups still occupy too small a place in the world. This seems responsible for our overall lack of sobriety, sanity, and serenity, the reason why peace on Earth does not yet prevail, and the reason why we are, to put it bluntly, using an American figure of speech, "headed to hell in a hand cart."[728]

It is important that we begin to use HPMAGs much more widely, to change our individual selves first, thereby greatly assisting the healing of the world's peoples in all walks of life. More sober, sane, serene, spiritual, ecological, moral, responsible, civil, free, and participatory, truly democratic citizens would not have to look backward with anger, fear, hatred, remorse, and resentment, but at themselves in the present, so that we might chart a brighter future, with all the faith, hope, charity, courage, respect, responsibility, and wisdom that we have at our command, leaving the remainder to the Higher Powers.

Alcoholics and their families resurrected Twelve Step Higher Powered Mutual Aid in the gruelling Great Depression of the last century. Today, Twelve Step HPMA is relatively successful, partway to blossoming worldwide.[729] While HPMA is not at all fully recognized for its enormous potential to work for the health of individuals, families, communities, and the planet, I am confident that this will change.

In the words of Georges Sioui:

> For human beings there is really only one way of looking at life on this earth, and that is as a sacred circle of relationships among all beings, whatever their form, and among all species. The great danger we face is that of reaching a point where we no longer see life as a vast system of kinship.[730]

---

[728] This adage is of unknown origin. One can replace "hand cart" with "hand basket."

[729] It would not be surprising if the Indigenous peoples - recovered somewhat from egregious persecution - have adopted a planet wide approach to HPMA as well, or are working toward it.

[730] Georges Sioui, *Huron-Wendat*, xi.

A reading list covering this subject would fill many library shelves. It would include Rachel Carson's *Silent Spring* (1962), Shepard and McKinley's *The Subversive Science* (1969), The Ecologist's *A Blueprint for Survival* (1972), The Club of Rome's *The Limits to Growth* (1972), Ward and Dubos's *Only One Earth* (1972), E. F. Schumacher's *Small is Beautiful* (1973), Willy Brandt's *North-South: A Programme for Survival* (1980), Capra and Spretnak's *Green Politics* (1984), Gro Harlem Brundtland's *Our Common Future* (1987), Murray Bookchin's *The Ecology of Freedom* (1991), Jack Weatherford's *Savages and Civilization: Who Will Survive?*, David Suzuki's *The Sacred Balance* (1997), Jared Diamond's *Collapse: How Societies Choose to Fail or Succeed* (2005), James Lovelock's *The Revenge of Gaia* (2006), Jeremy Rifkin's *The Empathic Civilization* (2009), Lester Brown's *World on the Edge* (2011), Parag Khanna's *How to Run the World* (2011), and Mary Ellen Abrams's *A God that Could Be Real* (2015). Why can we not hear these warnings and act accordingly?

Bill W. saw how we were filled with fear,[731] and how the world was coming apart at the seams.[732] From a Higher Powered Mutual Aid perspective, the problem may simply be that we do not truly know how to respect ourselves and others, to be dutifully self-regulatory, and self-responsible, to simply listen and speak humbly and honestly, and finally, with love in our hearts. HPMAGs can teach us a new global morality, to respect and to be responsible. I once heard the following, an old English proverb, from a Twelve Step friend:

> Fear knocked on the door.
> Faith answered it,
> And no one was there.

Keating wonders if we might replace faith in this proverb with love.[733] However, I believe that before we can love, fear must be overcome and

---

731 *Alcoholics Anonymous* (The Big Book), 67.
732 *Alcoholics Anonymous Comes of Age*, 233.
733 Thomas Keating, *Divine Therapy and Addiction*, 85-86.

trust must manifest itself, before we have faith. If we neglect to establish trust, and its follower faith, we fail to accept, or realize, that we must do the step-by-step footwork to get us to a firm foundation for love.

Our spiritual and ecological interdependency is being recognized more and more in the new millennium, not only by academics and scientists. A three-page November 2000 reprint of an August 1976 *A.A. Grapevine* article, titled "Don't Order Me Around," notes simply that A.A. is ecological. Much longer treatments of the history and meaning of HPMA have failed to recognize its holistic character, and its potential to bring us toward worldwide balance in all aspects of our lives.

Charity sometimes begins at home in HPMA. This is a charity rooted in true self-love, a compassionate charity in relation to others, and to the rest of Creation. Charity can be a dreadful word when used to deny others the opportunity to be or to become self-responsible. With self-responsibility comes true individual freedom, not the other way around. Individual freedom without self-responsibility is only a vortex of social and ecological contradiction and destruction.

As part of a conclusion in this context, we can refer again to Brian Murphy's book, *Transforming Ourselves, Transforming the World*. He observes: ". . . the precipitous withdrawal of the state and government from the historic responsibility of caring for its citizens and promoting the general welfare of society."[734]

Yet the original withdrawal of responsibility from citizens was a move away from an ecologically creative society toward an anti-ecological one, where citizens' self-responsibility was weakened beyond recognition. The violent construction of such artificial states, worldwide, has allowed authoritarian governments to replace benign, beneficial anarchy that mines the innate self-regulatory, self-responsible, and decision-making power of the "individual in relation." This has made it acceptable to abuse Nature and other beings, legitimizing the desecration of our Mother, the Earth.

---

734 Brian Murphy, *Transforming Ourselves, Transforming the World*, viii.

With that destruction reaching crisis proportions, the state seeks to download major responsibility to big business, rather than back to its citizens. Of course, big business and big government have long occupied the same bed, often meaning well. But if the current downloading continues, the path of ecological destruction would also continue. Why? Because this downloading does not begin with each individual's creativity and responsibility in relation to the other. Big companies and governments will necessarily fail, simply because they are not equipped for the job: *they are not "individual in relation" living citizens endowed with the innate creative power of unique, positive purpose(s).*

Peter Kropotkin grudgingly acknowledged that attempts at creating community failed because there was no adequate place for authority. Leaders were elected, not "trusted servants." (First Nations had practiced this for eons, and the Twelve Step pioneers came to know it well.) What could work much more widely, instead, is a benign anarchy of HPMAGs that place authority and faith in loving and benevolent Higher Powers. The result would be a healthy balance full of potential for imperfect individuals and groups who know that they don't have all the answers.

But not having all the answers - we are not God - and knowing that all the answers will never be available to us - certainly in my lifetime - is a critical realization, adding a lustrous gleam to the satisfaction and pride that we can have in just being human. It creates room for humility and serenity, but also for faith, hope, courage, and even passion.

The world has largely forgotten the spiritual and ecological principles of Higher Powered Mutual Aid, in a heady, lemming-like rush toward oblivion. I have cited some sources describing this tendency, but will not review the subject in detail. That is not my purpose. Yet neither will I offer fashionable, "positive" ideas about the special place occupied by humanity, next to God, or sitting childlike in Divinity's lap, waiting to take the throne should it be vacated. Given the "negative" views over the past forty years or more, some have capitalized upon, or created, a thirst for more positives.

One of the books putting this forward over thirty years ago was Marilyn Ferguson's *The Aquarian Conspiracy*. At one point, she argued for "... recognition that higher-order life is not bound by 'laws' but is capable of

On Evolution

boundless innovation and alternative realities."[735] While Ferguson did not define "higher-order life," there is no doubt that she is referring to human beings. Putting us at the pinnacle of an authoritarian, hierarchical creation, above so-called "lower-order life," is a dangerous undertaking, proven so now over 3000 years, particularly the past 5 centuries. Ferguson's work may have inspired hope for the Aquarian Age, but needed further grounding to be truly meaningful.

More recently, Neale Donald Walsch recorded a dialogue that he had with the God of his understanding. In *Conversations with God*,[736] God asks Walsch: *Can you conceive of yourself as one day being a God?*

Walsch replies, hopefully: *In my wildest moments.*

Then, predictably in a book written by Walsch, God encourages him, saying:

Good, for I tell you this: You are already a God. You simply do not know it. Have I not said 'Ye are Gods'?

This is not helpful. We are not God nor will we ever be.[737] The continued entertainment of blinkered "neo-ideas" about a limitless human future in a techno-wonderful Universe where we aspire to be God is just that: entertainment. It is not the serious undertaking of repairing, building, and sustaining a healthy world for as long as the sun shines. We have made amazing progress, moving from diminutive prey to the world's most fearsome predator at the top of the food chain. But we cannot eat our money, or the Earth, in order to survive. More humility, please.[738]

---

735 Marilyn Ferguson, *The Aquarian Conspiracy: Personal and Social Transformation in Our Time*, (Los Angeles: J.P. Tarcher, Inc., 1980), 167.

736 Neale Donald Walsch, *Conversations with God: An Uncommon Dialogue*, Book 1, (New York: G. P. Putnam's Sons, 1995), 202.

737 Benyus states that "Strapped to our juggernaut of technology, we fancy ourselves as gods, very far from home indeed." (Janine Benyus, *Biomimicry*, 1997, 5.)

738 Twelve Step writers can get carried away, too. Keating says that "You cannot do much better than to become God by participation." (*Divine Therapy and Addiction*, 106.)

James G. Duncan

In carrying the message, and in our efforts to change ourselves and change the world, living the Twelve Steps can be a humble example for others to follow. Ernest Kurtz et al devote a chapter to humility in the book, *The Spirituality of Imperfection*, a consideration of the spiritual findings of A.A. They suggest that an appropriate humility is not a "me last" put-down of the "individual in relation,"[739] and make further comments on destructive "me first" attitudes that scorn humility today. The authors capture the essence and importance of a balanced, honest humility.

We are each and all conceived and born of mothers and fathers "in relation" to one another, to the Earth, Water, and Sky, to all animate and inanimate Creation. There is, in truth, no way around this, and no amount of reasoning can undo such basic, knowable truths of our existence. The pursuit of individual creative self-interest will always be our primary motivation: we each begin in our own hearts, minds, and spirits. However, individual self-interest is circumscribed by the interests of the group, the family, the community, the world, and the cosmos: the common welfare comes first.

Yet, as earlier noted, by definition of our superbly ambiguous natures, the common welfare is itself dependent upon the creative contributions of safe, free, moral, self-regulating, self-responsible individuals. Each of us is, and will always be "in relation," and neither the individual nor society can thrive without a healthy relationship between them.

The next chapter builds upon the Twelfth Step possibilities of Twelve Step Higher Powered Mutual Aid as a way to help us change ourselves and change the world. This is something that the Twelve Step pioneers had foreseen, in their beliefs, words, and actions. Utilizing inspiration and knowledge as well from ancient, holistic, ecological First Nations Higher Powered Mutual Aid circles and societies, we will look at some specific, practical measures that we can undertake in what I call "Blue Green Politics."

---

739 Ernest Kurtz et al, *The Spirituality of Imperfection*, (New York: Bantam Books, 1992), 188.

CHAPTER 8

Blue Green Politics[740]

*Dakota children understand that we are
of the soil and the soil is of us...*

LUTHER STANDING BEAR

WE CAN BROACH the idea that much of the human population of the globe is ill in one way or another. Examples vary from mental, physical, and spiritual diseases, to greed, substance abuse, egotism and playing God, or using our great power to dominate and lay waste the rest of earthly Creation. Many of these illnesses are hidden from us because they are still seen as legitimate or "normal," such as externalizing pollution, a boundless love of money and possessions, or rampant "anything goes" morality. Wilson Schaef says that addiction is now the norm in society.[741]

The common denominator of all these illnesses, I believe, is our continuing futile attempt to break the ecological circle, something beyond our capabilities as Earth-bound creatures. We live on the Earth and are part of that circle/sphere, like it or not. The circle rules, quite untarnished

---

740 "Blue Green Politics" is used here rather than "green politics," since a blue green approach has the potential to move us toward a conservative (blue) self-responsibility. This shifts us away from the notion of the "free lunch" and religion, big government, or big business doing everything for us. Such a shift would allow us to take up our natural, and rightful, even dutiful, "individual in relation" responsibilities toward ourselves and the common welfare, around the world.

741 Anne Wilson Schaef, *Beyond Therapy, Beyond Science: A New Model for Healing the Whole Person*, 1992, 4-6.

by our efforts to break it with hammer or saw. Stepping outside the circle, as we have attempted to do in this short term, geologically-speaking, is the creation of illusion. Effectively, we live in trance. Nature does not work the way we would like it to, and will "throw us under the bus" if we continue our folly.

Of course, illnesses are best addressed when individuals recognize that they are sick. No one goes to a hospital emergency ward while feeling healthy, and no one goes to her or his first Higher Powered Mutual Aid meeting when entirely sober and sane. However, if we adhere to Scott Peck's view, that we are all emotionally ill to some degree,[742] few if any of us are sane, even if we do not recognize it.

Many see that their illness can be kept in check, and their sobriety, sanity, and serenity developed, by attending HPMA meetings regularly. At meetings, we come to see that the outside world is the really crazy place where little makes sense. I for one feel a certain warmth, like putting on a favourite sweater in wintertime, when I enter Twelve Step doors. Sometimes even the thought of going to a meeting is helpful and hopeful. It is often a relief after facing bizarre external society for a few 24-hour periods.

In this context, there are key questions that people might ask themselves. How are my relationships with other creatures, places, and things in the rest of Creation? Do I think, feel, and act as if the Earth is my actual home, or do I feel that it is more a landfill site, or a trap or prison that confines me? We are prisoners only of self-destructive conscious or unconscious emotions and behaviours. If we cannot free ourselves of these, they are our captors and we are their prisoners.

Extolling the virtues of space travel on a November 15, 2002 Canadian Broadcasting Corporation (CBC) television show, one rocket scientist said of our relationship with the Earth: "We are trapped here." Absorbed in his work, he may well have felt trapped, but I suggest that widespread acceptance of such ideas is a manifestation of disease and/or illusion. In its book, *Courage to Change*, Al-Anon offers the acceptance of human

---

742 M. Scott Peck, *The Road Less Traveled*, 17.

limitations as a way out: even if we were among rarified philosophers or scientists, there would be immeasurably more to learn. [743]

Once the choice is made to be well, much healing can take place in the Higher Powered Mutual Aid societies that already exist. The evidence presented earlier shows that A.A. and Al-Anon are relative newcomers to HPMA, but they have done near-miraculous work since their founding. Twelve Step HPMA is effective, not perfect, and it has the tools to be successful on a far broader scale than it has to date. The simple reason? Twelve Step HPMA is implicitly ecological, as First Nations HPMA is, explicitly so.

There is much room for expansion in the partnership between HPMA and the professions. Together they might reach toward the goal of healthier individuals, communities, countries, and a healthier planet. A.A. recognizes the importance of this in its pamphlet, "How AA members Cooperate with other Community Efforts to Help Alcoholics" (1974), and in "How A.A. Members Cooperate with Professionals" (revised 1994). The book *Al-Anon Faces Alcoholism* devotes half its space to specialists who work with the family disease of alcoholism.[744] However, while knowledgeable people do have an important role to play, Higher Powered Mutual Aid is much cheaper, and is sometimes more effective,[745] than professional health care, public or private.

Each and every "individual in relation" could develop a healthy attitude that recognizes the interrelatedness of all things and that there are no ecological shortcuts to happiness. This change of attitude is a first step in healing. Seen in this light, our first illness, now reaching pandemic

---

743 Al-Anon Family Groups, *Courage to Change: One Day at a Time in Al-Anon II,* (Virginia Beach, VA: AFG Headquarters, Inc., 1992), 225.

744 Al-Anon Family Groups, *Al-Anon Faces Alcoholism*, (Virginia Beach, VA: AFG Headquarters, Inc., 2nd edition, 1992).

745 Nowinski compared the A.A. model with other modes of healing, cognitive behaviour therapy and motivational enhancement therapy, and found Twelve Step assisted therapy just as effective, or moreso. (Joseph Nowinski, *If You Work It, It Works!* 2015).

proportions, was probably an emotional imbalance. Coercion is the culprit, fear, insecurity, and lack of trust the illness, and HPMA the antidote.

In my heart and mind, I know that any kind of reasonable, imperfect politics must begin with individuals who know themselves in relation to others, and whose centre of power resides within themselves and their Higher Powers. I am in the fullest agreement with people like the Twelve Step pioneers who saw that when people give up responsibility for themselves, their families and communities, to hierarchical human authority, they are giving up their most precious possession, and the most precious possession of any society: the God-given creativity of the individual functioning within the common welfare.

Based on what I have read, and the feel that I have been able to gain for the subject, it seems to me that the Twelve Step founders might have agreed with a "Blue Green Politics" for individuals fully practicing the principles of the program in all their affairs. Bill W. took the issue of energy conservation seriously in the early 1960s. *'PASS IT ON'* says that, though concern about the problem was still a decade away, Bill W. "... was, as usual, years ahead of his time."[746]

It is of no use to continue blaming those who have forcibly taken away, or who have blundered into taking away, the creative power of the "individual in relation." That barbarian, fear-induced, and fear-producing habit started long ago, and present-day martinets may see the madness of their ways through participation in HPMA. If hierarchical authoritarianism is seen for what it plainly is, perhaps the most severe illness, Controllers Anonymous (Con-Anon?) might become a force to be reckoned with.

A customary line has it that Twelve Step recovery is a process, not an event. HPMA takes time to work its results. Charity toward oneself is often the first concern. But just as importantly, charity can go some distance toward helping the less fortunate ill to regain a sense of themselves as being as full of positive purpose(s) as the more fortunate ill. "Blue Green Politics" may thus entail "Red Green" (liberal) measures for a time, so that the wider success of Twelve Step HPMA can include the less fortunate.

---

746 *'PASS IT ON': The story of Bill Wilson*, 380.

On Evolution

If people can't get to meetings, they can join "on-line" groups. HPMA meetings can be taken to them, too; this is already done to a degree in institutional Twelfth Step work.

Bill W. and the pioneers did not want government for Alcoholics Anonymous, and we know that, except in the short term, big government does not work in wider society. However, the divesting of government responsibilities needs to be accompanied, hand in hand, with individual and group investment in *self*-responsibility, something readily available for practice in Higher Powered Mutual Aid societies.

With "Blue Green Politics" there is no intention to bring business, government or religion to Higher Powered Mutual Aid. A "Blue Green" HPMA movement needs to be about healing from illnesses to be safe, free, creative, responsible, and participatory individuals, in ecological, democratic societies. In ancient First Nations societies, HPMA has long been a method used to maintain serenity, peace, and social order.

More recently with the Twelve Step movement, HPMA has become an implement for individual and social transformation, leading millions to sobriety, sanity, and serenity. Widespread application of a "Blue Green" variety of HPMA can lead us toward healthier, peaceful societies with ecological, economic, political, and social stability. Others are thinking along the same lines. Lachance feels that a Twelve Step program is needed to deal with environmental crises wrought by addictions in society.[747]

Once outside Twelve Step HPMA meetings, with recovery well on its way, individuals can enter a democratic arena where they have the responsibility to function as world citizens, as described by the Twelve Step founders. Outside meetings, Twelve Steppers can come together in groups to practice ecological "Blue Green" principles in the many other aspects of their lives.

The assistance of a wide and deeply implanted "Blue Green" HPMA movement, with circles and groups underlying democracy, could restore the power of the "individual in relation" to the broader economic,

---

[747] Albert Lachance, *Cultural Addiction: The Greenspirit Guide to Recovery*, (Berkeley, CA: North Atlantic Books, 2006).

environmental, political, and social stage. This "individual in relation" is a responsible and participatory citizen for whom true freedom and creativity can be gained only within the common welfare. The end result would be a "Blue Green" politics practiced by a critical mass of individuals in relation to all Creation.

Difficulties are bound to occur. Immersed in First Nations and Twelve Step philosophy, members can be circumspect, to see if outside groups are "joinable," that is, First Nations/Twelve Step friendly: the question might be, do they follow a form of HPMA already, or would they adopt HPMA as their own? That might occur only rarely, initially, suggesting that First Nations and Twelve Step members may need to first create their own "outside groups" to avoid the conflict and controversy prohibited in HPMA.

"Blues" (Conservatives) may favour Sky Gods, "Greens" often lean toward Earth Gods, and atheists and agnostics come in whatever door they wish. First Nations people choose what makes sense to them in Higher Powers, from Earth, Water, and Sky. Of course, there are many variations on the theme of God, numerous as the leaves it seems, but fortunately they can all be accommodated in the wide embrace of HPMA. The pioneers of Twelve Step HPMA ensured that their doors were wide enough to admit the non-believer as well as the believer, as in First Nations HPMA. Undoubtedly, one of the greatest aspects of HPMA is that it embraces *all* understandings of God, even the disbelief of atheists.

On a walk near my home in the early 2000s, I came across what looked like a new book, lying directly in my path. Inside the hardback cover, I found a handwritten number that I called on the phone; it was no longer connected to the system. The textbook, *Managing: A Contemporary Introduction*, had been published most recently in 1981,[748] but it had seen little apparent use. Two chapters seemed pertinent in my research: "Power and the Political World" and "Managerial Politics and Tactics."

---

748 Joseph Massie et al, *Managing: A Contemporary Introduction,* third edition, (Englewood Cliffs, NJ: Prentice–Hall Inc., 1981).

On Evolution

Delving into it more deeply, I found several descriptions of ethical conflicts in business. The most prominent of these was reference to a study of 3,000 U.S. businesspersons. It concluded that "...88% say a dynamic personality and the ability to sell yourself and your ideas is more of an attribute to the manager...than is a reputation for honesty or firm adherence to principles..."[749]

We have been bearing the fruit of such attitudes for centuries, with widespread fraud, insider trading, influence peddling, "Ponzi" schemes, and outright robbery. Suzuki notes that business schools still teach you to "leave your values at home when you go to work."[750] An October 19, 2015 TV Ontario program, *The Agenda*, noted that cheating by students was rampant; trained so, graduates may participate in unethical practices, without question. Al-Anon's *Discovering Choices* (2008) cites the words of one businessperson who said that he was rewarded at work for traits *opposite* to what he had learned in Al-Anon: he had to change or revise many of the self-serving principles that he used at work.[751]

Promisingly, the 1981 book also said, "...management in the 1980s and 1990s may become more oriented to the problems of the environment."[752] Although there has been some action on this front, it has not come about to a nearly sufficient degree, and the time is short to turn things around. In *The Revenge of Gaia*, Lovelock says that our war against the environment has been spreading with amazing speed, yielding alarming results.[753]

Mitroff et al use the Twelve Step model to propose a radical redesign of organizations, naming it "total ethical management," and call on business schools to change, too.[754] A ground-breaking book with a wealth of good ideas is *12 Step Wisdom at Work: Transforming Your*

---

749 Ibid, 427.

750 David Suzuki, *Good News for a Change*, 12.

751 Al-Anon Family Groups, *Discovering Choices*, 128.

752 Joseph Massie et al, *Managing: A Contemporary Introduction*, 2008, 468.

753 James Lovelock, *The Revenge of Gaia*, 2006.

754 Ian Mitroff et al, *Framebreak: The Radical Redesign of American Business*, (San Francisco: Jossey Bass Publishers, 1994), 138.

Life and Your Organization.[755] It explains how Twelve Step programs can help businesspersons to make their actions match their values. As in the Eighth of the Twelve Steps, where we make a list of those we have harmed, business organizations must do the same: take responsibility for their actions.[756]

Our present "*koyaanisqatsi*" (a Hopi word for crazy or insane life or "life out of balance")[757] culture can be changed if government and business would adhere to Twelve Step principles. Hammond's book explains how we might, through the application of these principles, move from an economic growth model with its sole monetary bottom line, toward sustainability and a commitment to the common welfare.[758] These are radical ideas,[759] the book says, requiring great courage to implement.

As part of a new beginning, we might have democratic governments, non-governmental organizations, big and small businesses, labour, religious, and other bodies, move in the direction of "Blue Green" politics defined in their mission statements and in how they function in practice. Encouraging a much larger participation of the citizenry of the globe in self-regulating, self-responsibility-inducing, freedom producing, ecological societies, circles, and groups, will help to shine the way with light. Choices remain.

Theory

My ideas on the theory of Blue Green politics are largely a summary of earlier thoughts put forward here and there in the main body of the text. More words are necessary to outline specific suggestions about ecological, economic, and political theory associated with a benign, beneficial anarchy

---

755 William Hammond III, ed., *12 Step Wisdom at Work: Transforming Your Life and Your Organization*, Hazelden Foundation. (Dover, NH: Kogan Page U.S., 2001).
756 Ibid, 38.
757 Ibid, 134. I had encountered these words before, at a movie called "K*oyaanisqatsi*."
758 Ibid, 215–294.
759 Ibid, 38.

of self-regulating, self-responsible citizens, groups, and circles underlying healthier democratic governments.

Benign and beneficial anarchy in HPMA is less about self-rule, here, than it is about the human *right* - latent within us - to be responsible for oneself, taking us all the way to the *duty* to be self-regulating and self-responsible, so that that right, and one's purpose and potential, are realized for our benefit and/or for the benefit of others in Creation. It is important to remember that beneficial anarchy is necessary in First Nations and Twelve Step HPMA societies, circles, and groups, and in other organizations that respect and promote the self-regulating, self-responsibility of every "individual in relation."

Al-Anon poses a relevant question for group discussion: Is there a line between helping others and robbing them of their chance to solve their own problems?[760] The answer to this depends on the seriousness of the situation. If it is a matter of life or death, one must take immediate action to help in any way possible, including personal intervention, or calling 911 for the police, fire fighters, and paramedics.

Otherwise, suggestive guidance can be used, such as information about Twelve Step and Distress centre help lines, to allow self-responsibility to develop without crossing that Twelve Step Al-Anon line. The Twelve Step phrase "Keep coming back" lets members develop their self-regulating self-responsibility, their innate power to solve their own problems.

Why might involvement in the benign anarchy of First Nations circles and Twelve Step groups have economic, ecological, personal, political, and scientific benefits for society? The answer is precisely this: the absence of rules in HPMAGs, without coercion, dominance, or hierarchy, is always in the context of the individual welfare and the common welfare, together, where, potentially, all may recreate themselves, all may participate, and all may benefit. The Garden is not so far away.

HPMAGs liberate the creative individual spirit within the common welfare, and could benefit democratic, ecological societies if implemented

---

760 *Many Voices, One Journey*, (Virginia Beach, VA: AFG Headquarters, 2011), 310.

well and on a much larger scale. The presence of deeply embedded and comprehensive Higher Powered Mutual Aid would help us to make democracy the best form of society, without qualification.

Georges Sioui offers some useful insights from historic First Nations cultures. He recounts Jesuit Lafitau's description of the contrast between coercive, authoritarian European political models, and those of a more gently persuasive First Nations model. Almost three centuries ago, Lafitau said:

> While the petty chiefs of the monarchical states have themselves borne on their subjects' shoulders and have many duties paid them, [First Nations leaders] have neither distinctive mark, nor crown nor sceptre . . . to differentiate them from the common people. Their power does not appear to have any trace of absolutism. It seems that they have no means of coercion to command obedience in case of resistance. They are obeyed, however, and command with authority; their commands, given as requests, and the obedience paid them, appear entirely free . . . Good order is kept by this means . . .[761]

"European chroniclers of all times" wondered how this could work.[762]

This brings directly to mind remarks made in 1946 by Bill W. about the skepticism of non-alcoholics over the functioning of an A.A. that had no human authority. Kurtz observed, for instance, that Bill W.'s psychiatrist, Dr. Tiebout, ". . . questioned . . . whether, given the self-embraced and self-defining 'anarchy' of Alcoholics Anonymous, the fellowship could ever hope to mature . . ."[763]

In *Twelve Steps and Twelve Traditions* Bill W. made a commentary similar to Sioui's, using his familiar method of rhetorical questioning. At one point, he asked what holds A.A. together, and answers that the suggested

---

761 Georges Sioui, *For an Amerindian Autohistory*, 1992, 48.
762 Ibid
763 Ernest Kurtz, *Not God: A History of Alcoholics Anonymous*, 1979, 128.

On Evolution

principles of the HPMA program need to be observed by the alcoholic if recovery is to proceed.[764] Lois W. cited this in her discussion of the Al-Anon Traditions, as "obedience to the unenforceable."[765]

This is almost exactly what Sioui quoted (above) about requests to obey leaders of First Nations HPMA societies, where "good order" was kept. A.A. has held together, growing to a worldwide membership of an estimated two million people. Al-Anon's figure is approximately 400,000, with a possible total of ten million together with other fellowships using the Twelve Step model.

In the larger scheme of things then, the Twelve Step movement is still quite small, a veneer of global goodness that is primarily North American. The message has not been carried nearly far enough yet, and there seems to be a sense of complacency, or, as already noted, a "quietism"[766] in A.A. and in other Twelve Step societies, working against further progress. There is, too, an undeserved skepticism about Twelve Step organizations in a world so wedded to the use of alcohol, other drugs, and to our overly tolerant "anything goes" culture. This is not helpful for those who need HPMA in their lives.

Plainly speaking, the anonymous Twelve Step programs are *too* anonymous, and it is sometimes difficult to find them. According to one alcoholic who knew Dr. Bob well, A.A.'s cofounder felt that we could break the anonymity tradition in two ways: by offering one's name to the press or radio; or by being so nameless that one can't be found by other alcoholics.[767] Likewise, First Nations HPMA circles are still difficult to find. I only learned about the First Nations Men's Healing Lodge, "Pinganodin," and some other First Nations circles/lodges, from a female friend of the "Fox Clan."

---

764 *Twelve Steps and Twelve Traditions*, 129-30.
765 *The Al-Anon Family Groups, Classic Edition*, 2000, 166. This is further noted in Al-Anon's *Many Voices, One Journey*, 2011, 386.
766 Matthew Raphael, *Bill W. and Mr. Wilson*, 55.
767 *Dr. Bob and the Good Oldtimers*, 264.

James G. Duncan

There is as yet little recognition that Higher Powered Mutual Aid can go far beyond healing and recovery work, into providing an Earth-altering replacement for humanity's suicidal course. It is difficult, too, for those who have not attended meetings and worked programs, to find out how HPMA works in practice. We can study and write about HPMA all day and talk about it all night, but only individual subjective experience, especially going to circle/group meetings,[768] can truly teach the freedom of self-regulating, self-responsibility that is available for individuals in relation in Higher Powered Mutual Aid.

With long participation in Twelve Step groups, and more recently in First Nations circles, I have shifted, politically, away from big government. The freedom and self-responsibility inherent in HPMA attracts me now to a "Blue Green" politics that employs:

- the free and creative "individual in relation" as its basic eco-political construct;
- the truth that the individual good is bound together with the common good, and;
- ecology, to allow politics and economics to recognize our interdependent natures.

If the individual and the community are interdependent, a specific logic is implied to describe the relationship between them: eco-logic, the logic of the home, of our Mother Earth ("eco" is from the Greek word *oikos*, meaning house or home). Thus, the individual and the community have an ecological relationship.

Other things follow, to name three:

- Individual safety and community safety are interdependent;
- Individual needs and community needs are interdependent, and;
- Individual health and community health are interdependent.

---

768 A.A. members sometimes say, keeping it simple: "Don't drink, go to meetings, and pray like hell."

On Evolution

The individual and the community need each other[769] and neither can survive and thrive without the other; their interdependence is factual and necessary.[770] We are innately individual and ecological, making give-and-take choices when working for the benefit of the individual good and for the benefit of the common good. In this context, we must bear in mind that, for individuals to thrive and for communities to hold together, the individual good must be good for the community, and vice-versa.

Ecologist Paul Ehrlich, with 30 books to his credit, feels that science must be assisted by quasi-religious movements to re-instill sustainable values.[771] Novak and O'Byrne search for new structures to assist ailing democracies, and write of the potential of religious, human rights, or environmental organizations to accomplish this.[772] In a 1997 book, *Community Development around the World*, Hubert Campfens sees mutual aid as a valuable tool for community development.[773]

Nancy Ryley refers to addictions on 38 pages in her book, *The Forsaken Garden*.[774] But she says, mistakenly, that "…the popular appetite for self-help (sic) therapies [e.g., Twelve Step] may increase rather than diminish our alienation from the environment."[775]

---

769 M. Scott Peck, *The Different Drum*, 56.

770 In Navajo society, "[A]n isolated or alienated individual is a sick one…" (Paula Gunn Allen, *The Sacred Hoop*, 88).

771 Paul Ehrlich, *The Machinery of Nature* (New York: Simon & Schuster, 1986), 17–18; Paul Ehrlich et al, *New World New Mind* (New York: Doubleday, 1989), 145, and; Paul Ehrlich et al, *Healing the Planet*, (New York: Addison-Wesley Publishing Co., Inc., 1991), 251–58.

772 Michael Novak, ed. *Democracy and Mediating Structures: A Theological Inquiry*, (Washington, D.C.: American Institute for Public Policy Research, 1980); Darren O'Byrne, *The Dimensions of Global Citizenship: Political Identity beyond the Nation-State*, (London, UK: Frank Cass and Co. Ltd., 2003).

773 Hubert Campfens, ed., *Community Development around the World: Practice, Theory, Research, Training*, (Toronto: University of Toronto Press, 1997), 31.

774 Nancy Ryley, *The Forsaken Garden*, 285.

775 Ibid, 195.

James G. Duncan

David Suzuki's 2002 *Good News for a Change* seems headed in the right direction, but the promotion of change by individuals in his "affinity groups"[776] appears misplaced. People within these groups may not have started with their individual selves as the necessary focus. Unfortunately, none of the authors in the last three paragraphs have examined the potential of First Nations or Twelve Step healing modes to attain their precious objectives: Higher Powered Mutual Aid is an answer to their quests.

Newsweek made prominent mention of A.A., Al-Anon, Alateen, and Alatot in its 1988 front cover story, an analysis of the growing phenomenon of "Children of Alcoholics."[777] Others sometimes have little to say about the Twelve Step movement. Only one mention is made of the vitality of the Twelve Steps in the January/February 1991 issue of the *Utne Reader*.[778] And the June 24, 1996 issue of *Time*, quotes numerous people on the subject of its cover story, "Faith and Healing," without mentioning First Nations or Twelve Step societies.[779] Short sightedness can be blamed, a perennial problem for those rushing about, seeking salvation in all the wrong places.

The fault may lay in both camps, in an underestimation in the "outside" world of the vast potential of HPMA, and in the still-too-secret, still-stigmatized anonymous societies, where details of the good news do not get out well enough. Again, might the suggestion that we make use of "attraction rather than promotion" in Twelve Step societies be reconsidered? A rigorous effort would be needed, starting with a single Twelve Step group conscience. In the short term, more promoters like Joseph Nowinski, Nancy Ellen Abrams, and William Hammond III, are needed to point us toward the Twelve Step HPMA way of life more effectively, one that leads to sobriety, sanity, and serenity.

---

776 David Suzuki, *Good News for a Change*, 80.
777 Charles Leerhsen et al, "Alcohol and the Family," *Newsweek,* January 18, 1988, 62-69.
778 Katy Butler, "Spirituality and Therapy," *Utne Reader*, no. 43 (Jan. /Feb. 1991): 77.
779 Claudia Wallis, "Faith and Healing," *Time*, June 24, 1996, 34-40.

On Evolution

As already noted, Nowinski entered a dispute with a science-based approach.[780] His use of "Making Alcoholics Anonymous Easier"[781] and others' use of "Twelve Step Facilitation"[782] thoroughly promote Twelve Step education among prospective members in the ways and means of A.A. fellowships that work. This approach lets A.A. mind its own business - helping alcoholics - without saddling it with misleading complications.

Twelve Step recovery work is difficult enough as it is. It *is* work after all, using "pick and shovel" - another A.A. aphorism - if necessary, to dig our way down to the truth. However, as noted, *outside* of A.A. meetings and those of other Twelve Step societies, members can - as full citizens - take on other aspects of living, including family, community, democracy, environmentalism, and human rights.

Practice

HPMAGs provide the means, as spiritual, ecological, non-governmental, mediating structures between individuals and larger society, to move us toward smaller, more effective, democratic governments. Here, people can also choose to be moral, spiritual, ecological, civil, creative, free, responsible, safe, sane, and serene, as well as sober, all within the bounds of the common welfare.

At the citizen or "individual in relation" level of "Blue Green" politics, there is much to be done in taking a thorough approach to all-important self-regulation and self-responsibility. Attaining these, in the end, is a key benefit to be won in HPMA, a benefit that we can practice morally, spiritually, ecologically, democratically, and serenely, "in all our affairs." We can cite a few areas for reflection.

Our addiction to oil can be described as "oil abuse"; we are dependent on it in copious amounts, in many ways, and replacing it will require much effort, imagination, and ingenuity. "Money abuse," too, is an illness

---

780 Joseph Nowinski, *If You Work It, It Works!* 2015.
781 Ibid, 54-58.
782 Ibid, 58-73.

with a very large negative impact on global society. Those addicted to oil, or to money, may choose Twelve Step HPMA to recover. Bill W. said that money could get in the way of living the A.A. program.[783]

Self-regulating, self-responsible citizens could turn these societal weaknesses to their advantage by taking ecological thinking and acting seriously, both for our children's benefit and for the benefit of the Earth. Our current economic and financial systems are not based on ecological principles,[784] and, in the short term, only patchwork or band-aid solutions keep them going. Plainly speaking, anti-ecological economics is not sustainable.

Economic and ecological breakdown will result from the use of simple economic but anti-ecological principles such as:

- Unfettered competition that encourages the spending of capital (e.g., natural resources), not the sustaining of this capital, and;
- The externalization of costs to others in God's Creation (e.g., pollution of the air, land, and water; road kill). This is the "free lunch" behind products whose costs are borne by other humans, other creatures, and the environment itself.

Counteracting current economic and financial models are many ideas from a responsible "blue green" perspective; books on the subject were written by Paul Hawken (1993), and by Hawken et al (1999).[785] The following are brief examples of the practice of Blue Green politics gleaned mainly from the social, economic, political, and ecological thought of others, such as the Green Party. The "Greens" first rose to national prominence

---

783 *Twelve Steps and Twelve Traditions*, 120–22.
784 As Suzuki puts it, "[I]f our economic system were…rooted in the laws of physics and thermodynamics, we wouldn't be having an environmental crisis now." (David Suzuki, *Good News for a Change*, 51).
785 Paul Hawken, *The Ecology of Commerce: A Declaration of Sustainability*, (New York: HarperCollins, 1993); Paul Hawken et al, *Natural Capitalism: Creating the Next Industrial Revolution*, (New York: Little, Brown & Company Ltd., 1999).

in West Germany, like the phoenix from the ashes of World War II, and its ideas are on the cutting edge of democratic politics in the world today. *Green Politics* by Capra et al, details some of the early history of the Green Party.[786]

Choices

Through personal, market, political, democratic, and ecological choices, a self-regulating, self-responsible citizenry would encourage mental, moral, physical, and spiritual wellness, preventive health care, and a blue green tax shift from incomes to pollution. A revenue-neutral removal of taxes on incomes, with more taxation on the use of oil, gas, coal, alcohol, tobacco, and illicit drugs, would decrease illnesses of many kinds, curtailing resource extraction and carbon emissions.[787] Better choices would include walking, running, cycling, in line skating, ice skating, skiing, surfing, snowboarding, public transit, and solar, wind, hydro, thermal, and hydrogen cell energy. This has previously been termed a "green" tax shift, but since it promotes a conservative (blue) self-regulating self-responsibility, and moves us away from big government and big business, a "blue green" tax shift is more appropriate.

Self-regulating, self-responsible citizens would make market choices that encourage fair trade, healthy lifestyles, and the production of locally grown foods. An end to the use of pesticides would allow us to eat only organic foods. Of course, this would lead to ending the production of all toxins currently dumped into the air, land, and water. With "individuals in relation" making self-responsible market, democratic, voluntarily simple choices, the economy itself would become ecological.

Road kill has become especially troublesome to me as a pedestrian and cyclist, in both city and country. I find about one dead animal every one

---

786 Fritjof Capra et al, *Green Politics*. (New York: E. P. Dutton Inc., 1984).
787 Hawken jests that "Planet Earth is having a once-in-a-billion-year carbon blowout sale, all fossil fuels priced to move, no reasonable offer refused." (Paul Hawken, *The Ecology of Commerce*, 1993, 84).

to two kilometres during bicycle trips in the countryside in the spring, summer, and fall. After consulting an on-line expert on the matter, we came up with a conservative estimate of three to four million dead animals on North American roads at any one time during the summer. If birds, earthworms, and insects are taken into account, this figure is a serious underestimation. In Canada alone, *70 million* birds are killed annually by various means.[788]

On-line figures also showed that motor vehicles were the leading cause of accidental deaths for children, mainly as pedestrians and cyclists.[789] From 1991 to 1995, *50,000* children died on U.S. and Canadian roads alone, 40 percent of the total for all countries in the Organization for Economic Cooperation and Development (OECD). Cavanagh et al tell us, "…more than 50 thousand people in the United States are killed in automobile accidents annually, with pedestrians counting for one-fifth of those."[790]

The entire issue of motorized vehicles - materials used in their construction, safety, speed, numbers, overuse, and power source - is a critical area for searching and fearless examination.[791] As an example, deciding not to build or purchase vehicles capable, except in emergencies, of travelling over 100-120 kilometres per hour would be a help.[792] Slower travelling vehicles save gas, another reason to keep speeds down.

Especially in urban areas, use of electric vehicles would be practical. Car and truck engines that turn off while waiting at stoplights save gas and pollute less. Self-responsibility would help us deal with these problems in a holistic manner so that the use of motorized vehicles is not bound up

---

788 CBC TV News, Environment Canada report, October 1, 2013.

789 UNICEF, "Child Deaths by Injury in Rich Nations," *Innocenti Report Card*, Issue Number 2 (Florence, Italy, February 2001).

790 John Cavanagh et al, eds., *Alternatives to Economic Globalization*, 2004, 184.

791 Cavanagh et al, eds., say that "…the private car uses more scarce resources in its construction than any other product on earth." (John Cavanagh et al, eds., 2004, 183).

792 It is incredible to me that automotive vehicles created now are capable of traveling at two times the 100-110 kilometres per hour upper speed limit.

On Evolution

with narcissistic "keep up with the Joneses" economic functioning. With more environmental awareness and higher fuel costs, solutions such as these are already in place.

A major drawback to air travel is the tremendous pollution that it causes. Remedies may be in the offing. The Saturday May 14, 2011 edition of the Toronto *Globe and Mail* referred to a solar powered aircraft developed in Europe (page A18). The Canadian Broadcasting Corporation also described a solar powered flight across the U.S., to be followed by a planned journey around the world.[793] Solar powered boats are being used, with a proposed trip, too, across global water bodies.[794]

Recent developments in the renewable energy sector indicate that it may be possible for us to switch to 100 percent alternative renewable energy sources (solar, hydro, wind, thermal, hydrogen cell) within thirty years. (See the 2011 documentary film, *The 4th Revolution: Energy Autonomy*). The producers foresee "the dawn of the solar age [and] a new beginning for everyone."

There are great benefits to be gained from living in a society that, through the *thorough* use of Higher Powered Mutual Aid, promotes self-responsible choices. We would help ourselves and others, decreasing the need for big government and big business. Self-responsible and self-regulating individuals would take care of themselves better, ecologically, mentally, morally, physically, and spiritually, lessening illness and the need for hospitals, doctors, nurses, and other caregivers.[795]

Similarly, self-responsible individuals and groups, adhering to HPMA principles, would govern themselves better around the world, morally, democratically, ecologically, and spiritually, greatly reducing criminality, thereby reducing the need for police, lawyers, forensic scientists, judges, juries, courts, jails, and prisons.

---

793 CBC TV News, June 16, 2013.

794 Royal Canadian Geographical Society, *Canadian Geographic Magazine*, June 2012, 54-55.

795 Health care is the largest consumer of Canadian tax dollars, contributing to deficits and debt.

James G. Duncan

"CRADLE-TO-GRAVE" PRODUCT RESPONSIBILITY

Here, self-responsible manufacturers and consumers simply guide the ultimate return of products to the Earth in forms that the Earth can readily use once again. Organic shopping bags and car parts have made an appearance. Hawken expands the idea to "cradle-to-cradle" manufacturing, to encompass more effective recycling.[796] Cavanagh quotes Fritjof Capra as naming this a "closed loop system" where "…the waste of any one organization would be a resource of another."[797] Lulic calls this the "responsible creation" of products.[798] To reduce the amount of waste we produce before such products are widely available, self-responsible citizens might support policies for comprehensive recycling and composting, and for entrepreneurs to make profits at dumpsites.

ECOLOGICAL MORAL INVESTMENT

"Ethical investment" is a popular way to express the wise use of money, to try to encourage people to be more ethical, but it often leaves ethics too open to question. Ethics can vary from person to person, culture to culture, country to country. "Ecological moral investment" is an all-inclusive package, without loopholes. Investment in companies and organizations with "cradle-to-grave" or "cradle-to-cradle" circular responsibility for products that the Earth can readily re-use, are examples of ecological moral investment.

Self-responsible, self-regulating citizens, investors, businesses, and politicians would ensure, through democratic means that ecological, moral safeguards are in place worldwide. In such a scenario, it would be impossible to "externalize" costs to the environment, to other countries, or to other beings. The same would apply to outer space, where we already see the return on our investment to the Earth in the form of space junk.

---

796 Paul Hawken, *The Ecology of Commerce*, 1993, 71.
797 John Cavanagh et al, eds., *Alternatives to Economic Globalization*, 2004, 195.
798 William Hammond III, ed., *12 Step Wisdom at Work*, 2001, 287.

On Evolution

FAIR MARKET PRICING
More ecological moral investment might be directed toward fair trade, to allow all producers and consumers to reach toward their innate potential for self-regulation and self-responsibility. There is much potential for good from the use of this type of self-responsibility-enhancing policy. If enough consumers decide that they want fair market products that are good for their sisters and brothers - and for the rest of Creation - economic theory suggests that we would have a new moral and ecological world. Morally and ecologically priced goods are responsibly priced goods.

ALIGNMENT OF ACCOUNTING AND ECOLOGICAL PRINCIPLES
The shortest route to practical and effective blue green policies may simply be the alignment of accounting and ecological principles. It has never ceased to amaze me that we have been able to carry out, through the legitimacy of human law, what is politely known as the "externalization" of costs. The clearest example of this is air, land, and water pollution, in which politicians, businesses, and the public have come to believe that sending pollutants to the environment is legitimate, that polluting (i.e., poisoning) the planet, our home, is a benign part of living, working, and making profits.

Dumping sewage into waters that we and others drink takes us beyond stupidity into madness. The operating motto, sometimes used by misled parents for their children, is therefore "out of sight, out of mind." A major theme in Hawken's *The Ecology of Commerce*, is the commercial and industrial *internalizing* of all costs, full cost accounting, so that externalities would cease to exist.[799]

The problem becomes all the more invisible when people pay to send their garbage to other locations. Toronto shipped its waste to Michigan. Hardly benign, this externalization of costs is really a euphemism for the destruction of our home, and our descendants' home, the Earth. A self-regulated and self-responsible citizenry can help to make that clear through ecological, economic, moral, and scientific, and social choices.

---

799 Paul Hawken, *The Ecology of Commerce*, 1993, 75-90.

The effective use of policy tools, taxation, accounting or other practical measures to help create healthier communities and a healthier planet, would be preceded and underlain by extensive use of Higher Powered Mutual Aid, for the re-creation of the moral, sober, sane, and serene "individual in relation."

The theory and practice of blue green politics requires wide discussion in groups outside of First Nations and Twelve Step meetings, groups that can be formed to practice HPMA principles in all their affairs. It must make sense to people who, ultimately, make the relevant choices. Just as real lasting peace, or serenity, cannot be imposed from the outside, a thoroughgoing blue green politics also begins as an "inside job" by serene individuals in relation to God's Creation.

A few more words are necessary to put this brief discussion of blue green politics into the context of the preceding chapters on Higher Powered Mutual Aid, particularly in the Twelve Step movement that I know best.

HPMA and Democracy: One Step Ahead

Before Dr. Bob passed away in 1950, Bill W. gained his agreement to strengthen the benign anarchy of A.A. groups with a democratic structure, the General Service Conference (GSC).[800] The GSC is attended annually by elected A.A. representatives from across North America. Al-Anon has a similar structure, the World Service Conference (WSC). These bodies are guided by the Twelve Steps, the Twelve Traditions, and the Twelve Concepts, so that service conference representatives are the voice of Twelve Step groups, and not in authority over these groups.

As are all Twelve Step leaders, service conference representatives are "trusted servants; they do not govern."[801] This is true democracy in practice, not only in word, now in operation for many decades in A.A. and Al-Anon. It is undoubtedly among the finest results of the leadership

---

800 *Dr. Bob and the Good Oldtimers*, 1980, 325
801 See A.A.'s Tradition Two at the front of the book.

of the Twelve Step pioneers, with enormous remaining potential for good in the world. Regrettably, it is still a secret too well hidden from those searching for better modes of human organization. Might we look beyond, e.g., the stigmatization of Twelve Step societies, or beyond the racism that still occurs against First Nations peoples, blacks, and others? We can only hope and work toward such goals.

The Twelve Step founders never leaned toward socialism. The systematic, hierarchical politics of the removal of individual self-regulating self-responsibility, went against their beliefs, as it did for Kropotkin. Instead, those pioneers produced a deceptively simple, solid theoretical base, and practical structures for recovery from alcoholism, with much greater promise for the "myriad ills" of families, communities, and the planet. The Twelve Step founders deliberately devised organic anarchic/democratic fellowships, guided by Higher Authority, rather than the inorganic, toxic hierarchical societies that we have today, primarily serving greed.

"[E]ach of A.A.'s principles," Bill W. wrote, "*every one of them*, has been borrowed from ancient sources."[802] The First Nations of the Americas, and no doubt elsewhere around the world amongst other Indigenous peoples, have long known the value of placing authority in Powers greater and wiser than themselves. Higher Powered Mutual Aid was a way of life for First Nations peoples to the extent that the guidance of Higher Powers was implicit in all that they did. It is no accident that there are strong parallels with their younger siblings, the Twelve Step societies.

Peter Kropotkin's lack of a secure place for authority, beyond the reach of the grasping "power driver," as Bill W. sometimes described himself,[803] was a large contributing factor in the failure of mutual aid societies of Kropotkin's time. His writings omit spirituality and religion from serious consideration. Kropotkin's belief in science, economics, and ethics, but not in a central Higher Authority, doomed his efforts to failure.

It is important to note again that Twelve Step societies were brought into being in a democratic country where the membership was relatively

---

802 *Alcoholics Anonymous Comes of Age*, 231.
803 *'PASS IT ON': The story of Bill Wilson*, 156.

free to pursue peaceful individual recovery within the common welfare, without having to deal with overwhelming outside issues that might impede them. Yet, even with great success, the Twelve Step pioneers were under no illusion that they had built the perfect society. In 1957, Bill W. observed that there were serious outside dangers that ". . . could prove to be our undoing . . . stronger societies than our own have been undone."[804]

Worldwide economic difficulties, growing environmental problems, and international conflicts, including ethnic cleansing and refugee crises, are but a few examples of how Higher Powered Mutual Aid circles and groups might be affected in negative ways. Conversely, the much wider use of HPMAGs in the world would be an effective means of dealing with such crises.

The essential removal of human authority over others necessitates its placement outside ourselves, in Powers greater and wiser, in order to keep humanity together in humble and common cause. In this, the Twelve Step pioneers followed William James and Carl Jung, rather than Peter Kropotkin.

---

804 *Alcoholics Anonymous Comes of Age*, 233.

CHAPTER 9

Conclusions

With all its sham, drudgery, & broken dreams,
it is still a beautiful world.

MAX EHRMAN

AN OVERALL GOAL of this book was to add suggestions about Higher Powered Mutual Aid to the mix of ideas on current local and planetary predicaments. HPMA was nearly abandoned before and during the expansionary years of the Europeans, and beyond nearly to the present time, and the results have been disastrous on a global scale. But the struggles and successes of Twelve Step fellowships, and, more lately, of now-renewed First Nations societies, have made the benefits of HPMA recognizable once again. The much greater success of Higher Powered Mutual Aid, and of humanity, I believe, awaits the involvement of far, far more of the world's citizenry.

As the Twelve Step pioneers foresaw, and as I have outlined in brief, the strengths of Higher Powered Mutual Aid, exemplified in First Nations circles and Twelve Step groups, might be used to help rebuild our societies from the bottom up. Robust HPMA societies could express a return to progressive human evolution in cooperative, moral, scientific, spiritual, democratic, and ecological HPMAGs, where both the individual good and the common welfare can be attained, together. Higher Powered Mutual Aid is the far-underrated and long-ignored factor of evolution and Creation.

At the beginning of the book, I set three objectives. The first was to trace the history of Higher Powered Mutual Aid in the Twelve Step movement and its use of ideas from people like Jung, James, and Kropotkin. Starting in the Great Depression of the 1930s, alcoholics like Bill W., Dr. Bob, and their spouses Lois W. and Anne S., all discovered that egalitarian non-authoritarian mutual aid was a fruitful path for them to follow, yet only made workable with the guidance of benevolent Higher Powers.

After the publication of the Big Book in 1939, some of Kropotkin's ideas on mutual aid were used to add to the experience of A.A. groups in formulating the Twelve Traditions. Critically, the pioneers "stepped up" their program, and did Kropotkin and others one better. They made mutual aid Higher Powered, and that has made all the difference. Placing authority beyond the human sphere, in Higher Authority, as it has long been in First Nations societies, is a key to dealing with our seeming helplessness.

In wider society, too, over the past five centuries in particular, it has become evident that we cannot handle authority by ourselves. From a holistic First Nations and Twelve Step perspective, how it works is simply through the placement of authority in the hands of Higher Powers, while doing our best to use humanly knowable truth to promote individual freedom within the common welfare. Jung's and James's powerful spiritual insights inspired A.A. to take the steps needed to surpass Kropotkin's form of mutual aid.

The second objective was to describe the theoretical and practical advantages of adopting the ecological Higher Powered Mutual Aid that is already available in First Nations and Twelve Step societies. This was done in Chapters Four through Seven, especially in the emphasis on the Twelfth Step practice of principles "in all our affairs." Included were significant parallels between First Nations circles and Twelve Step groups, as the basic building blocks of their respective societies.

The role of Higher Powers is central to First Nations societies, and there is a clear responsibility for First Nations peoples to practice their principles in all aspects of life, explicitly embracing all of Creation. Twelve Step groups and members have a less well-defined, but still explicit responsibility to practice the principles of Twelve Step Higher Powered Mutual Aid in every aspect of their lives.

On Evolution

The "kinship" or interdependent state of being nurtured in all HPMAGs is the base from which we can learn the further principles of self-regulation, self-responsibility, and self-respect. By ridding ourselves of authoritarian hierarchy, by learning HPMA principles and practicing them in all our affairs, we can each begin to govern our individual lives "in relation," in healthier ways, in our families, communities, countries, and upon the planet.

The book's third objective was to make broad suggestions for the furtherance of Higher Powered Mutual Aid to move us toward individual and group self-regulation and self-responsibility, in more effective, moral, and representative ecological democracies. A necessary start is to carry the message to *billions*[805] more "individuals in relation," to enable them to participate in HPMA circles and groups, in all aspects of living. This, I am convinced, would work miracles in this troubled world.

I have tried to show that, in strengthening ideas about mutual aid, the Twelve Step pioneers were able to tap into the creativity defined by the "individual in relation" to the common welfare. In the context of the third objective, perhaps most important is the recognition of similarities between the older traditions of First Nations and more recent Twelve Step societies, in their reaching toward the individual good and the common welfare, together. A melding of these traditions is needed, producing openings for leadership from both kinds of HPMA. It inspires me that Higher Powered Mutual Aid, while already demonstrating some success, has so much remaining potential for good in a humanity leaning towards an early end.

That potential might blossom more fully if enough people carried the message to others, who can then decide to change, to become well, by joining and participating in HPMA circles and groups. If that were to happen in sufficient numbers, why then, by definition, the battle would be won without firing a shot. Worldwide recognition of HPMA as a means

---

[805] We saw that Browne-Miller has estimated that there are over two billion alcohol and drug users in the world, together with smokers, gamblers, sex-addicts, and on and on. (Angela Browne-Miller, xxv-xxvi).

of getting us back to the Garden, and a federation of HPMA societies, of sisters and brothers, is what we need to break free of the chains that bind us. We can then enter a New Age of equality, freedom, humility, justice, morality, peace, self-responsibility, sobriety, sanity, and serenity, under God as we each understand God to be.

Practically speaking, it might not be possible to explore such a federation of HPMA societies immediately. We may not have the time, at present, to adopt such progressive worldviews as would be provided by the numerous First Nations societies. First Nations approaches to spirituality, for example, vary from First Nation to First Nation, almost from Elder to Elder, and can be complex, as is Nature itself.

Nevertheless, First Nations peoples can unite, if necessary, to work for the common good. There is in fact a long history of intertribal cooperation, taking us back at least to Shawnee leader, Tecumseh, who campaigned to unite First Nations opposition to the westward expansion of the United States in the early 1800s.[806] Black Elk, second cousin to Crazy Horse, tells of cooperation between his Oglala Sioux and other tribes, all fighting this westward expansion in the mid to late 1800s.[807] Canada's westward expansion took place with explorers, fur traders, the Royal Canadian Mounted Police, the hanging of Metis leader, Louis Riel, and a national railway that rammed through First Nations territory.

In 1992, Indigenous peoples gathered in Canada to observe the 500th anniversary of Columbus's "discovery" of the Americas. Current attempts are being made by the "Idle No More" movement in combating racism and struggling against the abrogation of treaties. Proposals for oil pipelines without proper consultation are being opposed by First Nations and others. So, despite my hesitation, cooperation between First Nations and other Indigenous peoples might allow them to take a lead in promoting Higher Powered Mutual Aid.

---

806 Nabokov, *Native American Testimony*, 95-98.
807 Black Elk, *Black Elk Speaks*, as told through John G. Neihardt, (Lincoln, NE: University of Nebraska Press, 1979), 10, 81, 93, 95, 134.

On Evolution

Encouragingly, some First Nations spiritual practices can be learned and adopted widely, in order to bring Twelve Step HPMA into closer communion with the Earth, Water, and Sky that are, of course, included "in all our affairs." The holding of hands in circles at the end of meetings - a practice in both First Nations and Twelve Step HPMA - is an example. In one simple Mohawk practice, the left palm faces up in the circle and the right faces down. Explained one Elder:

> The representation of the left palm [facing upward] in relation to the Earth is that our heart is on the left side of our body. Since we associate our heart as the source of loving, it becomes a metaphor of the love and generosity of the Earth in supporting all life. The skyworld bodies in our system are in sync [represented by the right palm facing downward] with the life force here on Earth. Hence bringing them together reminds us of our connection with all life. Also, our original instruction was to give thanks daily for these gifts.

Spiritual but not religious Twelve Step programs are relatively straightforward, step-by-step programs of reflection and action, ready-made, and adaptable for use by those of any religious faith or of none, including those of First Nations persuasions.[808] In the short term, we might continue along with this simple program to awaken our senses to the wider world and to the further potential of Higher Powered Mutual Aid. Subsequently, we might take on the challenges of ecological First Nations HPMA that promises deeper knowledge and understanding of the world we live in.

For now, Twelve Step societies provide a spiritual path that cuts through the Gordian knot of religious complication. They offer a new way of living that we can follow in all our affairs, a crowning achievement of the pioneers. Humanity is confronted by an array of serious problems

---

808 Keating is one of the latest to recognize this, saying: "[E]verybody can benefit from . . . the Twelve Steps . . . no matter what their [religion]." Thomas Keating, *Divine Therapy and Addiction*, 111.

that threaten its very survival, but HPMA programs provide us with a way out. To come to its senses, the world needs to give First Nations and Twelve Step Higher Powered Mutual Aid a fair trial, for as long as it takes.

There does seem to be a certain desperation now, in the air, fostering books such as Khanna's *How to Run the World*[809] and perhaps my own. Khanna feels that all the world's problems can be solved through better diplomacy, and, somehow, with "everyone watching everyone at the same time."[810] He cites the success of micro-finance, a "bottom up" small business practice that puts individuals in charge of their own God-given creativity and self-regulation, without the need of hierarchical, authoritarian government or big business.

It is important to restate the differences between First Nations and Twelve Step HPMA, and the mutual aid envisioned by Peter Kropotkin. The first and foremost difference is in the placement of authority. Twelve Step societies recognize benevolent, guiding Higher Powers, placing authority beyond the reach of those inclined to dominate, and so diminish individual freedom and creativity within the common welfare.

Kropotkin, however, found that placing authority in human hands was very disappointing to him. As noted in the introductory chapter, he admitted that:

> In the hundreds of histories of communities which I have had the opportunity to read, I always saw that the introduction of any sort of elected authority has always been, without one single exception, the point which the community stranded upon . . . [811]

---

809 Parag Khanna, *How to Run the World: Charting a Course to the Next Renaissance*. (New York: Random House, 2011).

810 CBC Radio One interview, 9:00 a.m., March 4, 2011.

811 Peter Kropotkin, "Proposed Communist Settlement: A New Colony for Tyneside or Wearside," in *Small Communal Experiments and Why They Fail*, (Petersham, Australia: Jura Books, 1895, republished 1997, 16).

On Evolution

Placing human authority in the hands of Higher Authority sets up the essential partnership between humans - and that which can be accomplished as humans - and Higher Powers, whom we must trust to do the rest for us, come hell or high water. Human authority over God's Creation, in authoritarian hierarchy, was, and still is, a societal structure built for war, not peace. Founded in anger, fear, and greed, authoritarian hierarchy has only created more of the same,[812] finally producing the ecological war to end all wars against ourselves and the rest of Creation.

The second essential difference between HPMA and Kropotkin's mutual aid is disease itself. One might characterize the illness of the global population today as resulting from the loss of the "individual in relation" to the rest of Creation. We have forgotten or set aside the fact that we are humble social and ecological creatures, as are all. HPMA recognizes this in its emphasis on healing the "individual in relation" in benign anarchic/democratic groups, rather than in authoritarian hierarchical groups. Kropotkin explained that mutual aid is a larger factor of evolution than competition, to counter Darwin's "struggle for existence" and appropriated "survival of the fittest."

However, Kropotkin was unable to deal effectively with the dangerous authoritarian diseases of his time, partly because he did not see authoritarianism as a disease. Higher Powered Mutual Aid is a healthier factor of evolution for ourselves, our communities, and the planet, since it fully enlists the support and guidance of Higher Authority. Because HPMA societies remove human authority from consideration, they are much more helpful than simple mutual aid. "Individuals in relation" can then practice unfettered self-regulation and self-responsibility, with the right - and, again, duty - to assiduously reach toward one's potential and purpose within the common welfare.

In spite of huge sums spent, diseases remain a threat. Authoritarianism is a disease. War and ethnic cleansing flourish. Alcoholism and other addictions are rampant. The Ebola, HIV, and Zika viruses are dangerous.

---

812 Charlotte Kasl touches upon this in her *Many Roads, One Journey: Moving beyond the Twelve Steps* (New York: HarperCollins Publishers, 1992, 53).

Antibiotics lose effectiveness. Dutch elm disease, the emerald ash borer, and the mountain pine beetle have decimated our forests, leaving them susceptible to uncontrollable fires. Fresh waters are crowded with invasive species and pollutants, such as tiny particles of plastic that find their way from domestic washers into the salt waters of the Earth.

The worldwide transformation of people, other life forms, and natural resources into money is, upon scrutiny, like alcoholism, a deadly illness. Paul Hawken says, "[C]ompelling evidence suggests that the behavior of many individuals in the modern corporation is remarkably similar to that of addicts."[813] Brain scans show close similarities between drug addicts and those addicted to investing with money.[814]

Kropotkin's countryman and contemporary, Leo Tolstoy, wrote of parallels between drinking and spending money in *Anna Karenina*. This was brought to my attention by a marvellous Al-Anon friend. Near the book's end, Tolstoy's character Levin had recently married and was spending time in Moscow; Tolstoy wrote:

> It was only during the first few days in Moscow that Levin had shown any astonishment at the unproductive but inevitable expenditure which seems so strange to one who lives in the country and which was demanded of him on every side. Now he was used to it. In this respect the thing that is said to happen to drunkards happened to him too; the first glass sticks in the gizzard, the second flies down like a buzzard, but after the third all go down like little birds. When Levin had changed his first hundred-ruble note in order to buy liveries for his footman and hall porter . . . this hundred-ruble note still stuck in his gizzard. But the next . . . went much more easily than the first. Now, however, the notes he changed . . . flew away like little birds.[815]

---

813 Paul Hawken, *The Ecology of Commerce*, 123.

814 CBC Radio One, April 9, 2009.

815 Leo Tolstoy, *Anna Karenina* (New York: Signet Classic, first published 1877, 1961), 671-72.

On Evolution

The Earth is a temple of Higher Powers, as well as our home, and anti-ecological addictions cannot prevail. If addictions win, we lose, all of us.

It is essential to restate, too, that the incorporation of the benefits of involvement in the benign anarchy and democracy of First Nations and Twelve Step HPMA societies, can only be gained through subjective experience, especially attending circle/group meetings. The atmosphere of recovery from disease in the benign anarchy of natural HPMA circles and groups, without rules or human authority, is very different from the atmosphere in hierarchical societies with human-created rules.[816] This is not immediately apparent when we have lived our entire lives under authoritarian hierarchy. The latter societal model cannot provide the tools to change ourselves and change the world for the better. It preoccupies us so much that we have little time to see other ways of living.

As First Nations HPMA has always done, Twelve Step HPMA gives us a set of tools, among them the Steps and Traditions, group guidelines, literature, slogans, sponsors, and of course Higher Powers. Through their use, we find a new way of understanding ourselves and all our relations, at meetings and in all our affairs. We learn to live freely within the common welfare, under guiding Higher Authorities that like the way we live.

It is entirely possible, still, to heal from illnesses that have resulted from the loss of the "individual in relation." However, the first task for individuals is to choose to return to the HPMA group, the circle, to learn its lessons. Already we have numerous community circles that enclose us (at home, at work, at church, in sport . . .), circles that are really strands of the spherical web of mutual aid, like the Earth itself.

Following from the thought of the Kropotkinist mutual aid evolutionists, and from First Nations and Twelve Step thinking, the imperfect individual good plus the imperfect common good equals the imperfect planetary good. This is an example of imperfect planetary or ecological reasoning. The Earth is not a perfect sphere and we are not

---

816 One of the greatest problems with authoritarian hierarchy is that it creates "experts" at the top, always ready to dispense what we are to think, believe, and act, removing the inner power of self-investigation.

perfect beings. Explaining this can be simple, like the child's game of connecting the dots, or molecules, if one wishes to be scientific. Our individual molecules (dots) are connected by a line or lines of molecules of the Earth, to people on other continents or islands, or to our next-door neighbours. Similar lines can be drawn through water or air.

As earlier noted, molecules are physically supported by, and are supportive of, other molecules - mutual aid or mutual support - in air, land, water, vegetation, lava, humans, other animals, and all the animate and inanimate. If molecules did not assist each other in this way, the world, and perhaps the Universe, would collapse in a moment. There are countless examples of mutual aid in a natural world where it is by far the largest factor of evolution. Darwinian competition here is insignificant.

Katz et al commented, 40 years ago, on the "erosion of the family structure" in the face of widespread socio-economic breakdown. They were in essential agreement with Kropotkin.[817] But so-called individuality has led us more deeply toward *ecological* erosion and breakdown. Mutual aid, seen by Katz et al as "spontaneous" in origin,[818] is, moreover, the evolutionary building block of any society, naturally innate in the molecule, the ant, the plant, and the human being.

Each human being needs to first undertake a thoroughgoing Twelve Step program whereby we can bring a Higher Power of some description into our lives; it does not need to be "God." We must make a moral inventory; admit our errors to self, other, and Higher Powers; make amends for wrongdoings; pray and meditate; carry the message to others; and bring the principles into effect in all our affairs.[819] This is nothing less, I suggest, than essential business for us as "individuals in relation," the world over.

Using our program principles to apply ourselves to the findings of science is a next step, perhaps at first glance "eco-alignment." In the

---

817 Alfred Katz et al, *The Strength in Us: Self-help Groups in the Modern World*, (New York: Franklin Watts, 1976), 4.

818 Ibid, 9.

819 See A.A.'s Twelve Steps at the beginning of this volume.

On Evolution

context of the book, this is a melding of First Nations and Twelve Step HPMA, ecological morality, and environmental science,[820] bringing us back to the Garden where we can be as one. One First Nations Elder lauded Canada's David Suzuki for bringing Nature to the television screen.

Altogether, we have to be careful, ever-watchful, and vigilant for the tendency to consider ourselves superior to others in every way. It must be stressed, again, that we are not God, nor will we ever be. We must only align ourselves with Nature, and with Higher Powers, through prayer and meditation, improved morality, right intention, thinking, reflection, and action.

The web of mutual aid and competition is hard-wired in each being, "individuals in relation," giving us the choice to participate or not to participate. Whether we are conscious of it or not, this natural, ecological web cannot be destroyed. That is just the way life is, the way we are, have always been, and will always be.

To continue to live as though the web can be done away with, like a toy, is only a fool's game. We have chosen to reside outside the web, partly unconsciously, a position from where we can now recognize this consciously. If we stay the same course, we will thereby be excluded from it by Powers greater and wiser. Change will come anyway, and we can make the best of it, not the worst, with the practice of HPMA.

Choosing to step outside the circle, the spherical web, as we have done, is the creation of protective illusion. Rectangular maps on walls, dissecting the living planet, hide the truth that everything is interconnected. Our resulting perception of the world, seen as through a filter or glass, allows us to live in another-world trance, to protect ourselves from fear and truth after thousands of years of warfare against ourselves and the rest of Creation.

Having had the privilege of practicing Higher Powered Mutual Aid for a number of twenty-four hour periods, I feel I may now carry the message to others in a reasonably humble manner. If I have been able to learn well

---

820 While riding Ottawa's O-Train one evening in 2014, I conversed with a young student who told me, with some enthusiasm, that he was practicing evolution in his laboratory!

enough the logic of the home, eco-logic, in my new home, HPMA, then I can continue to practice that same logic in all my affairs. I can rejoin the web of society, to help heal the larger global web, too, so that it can function better for my benefit and for the benefit of all.

However, if wider human society does not function well enough in the web - and it currently does not - then it is of no Earthly good to me, and I might not survive, just as surely as none of us will. If it is true that the meek will inherit the Earth, they might be from another species that work better together, like insects or birds. Yet, with the modest optimism born of my new way of life, I believe that Bill W. was right when he mused about working the Twelve Steps, saying that the only pressing issue is to make a beginning and persevere in our efforts.[821]

We have made the beginning, thanks to First Nations and Twelve Step societies. Let us keep trying, carrying the message to others and practicing the HPMA principles in all our affairs, with the hope and promise that each new day brings, while bearing in mind Alan Paton's humble and enigmatic words about expectations:

> [W]hen that dawn will come, of our emancipation, from the fear of bondage and the bondage of fear, why, that is a secret.[822]

---

821 Alcoholics Anonymous, *Twelve Steps and Twelve Traditions*, 68.
822 Alan Paton, *Cry, The Beloved Country*, (New York: Scribner, 2003; originally published 1948), 312.

Epilogue

Offering up a few gems in his 1994 book, *Savages and Civilization: Who Will Survive?* Weatherford wrote the following:

> During the twentieth century, civilization experienced a number of major scares…warning shots. Civilization proved capable of waging war on itself…we developed nuclear energy, and came close to provoking a nuclear holocaust, and we do so yet. When we survived World War I…World War II, and…the nuclear threat of the Cold War, we felt safe. When catastrophe did not follow…we felt relief, as though the danger had passed, but danger still approaches us.[823]

Weatherford is accurate when he says that we have proven capable of waging war on ourselves. This may be self-fear, elevated to self-hatred, born only of our inability to be God and to do away with all our faults, weaknesses, and pain. The plague of wanting everything, to go everywhere, and to be everyone but ourselves, has led us far astray.

Tolstoy wrote, in *Anna Karenina*, another apt slice of life in a restaurant conversation between friends Levin and Oblonsky:

> "You don't care much for oysters, do you?" said Oblonsky, draining his glass. "Or are you worried about something? Eh?"
>
> He would have liked Levin to be in good spirits, but Levin, though not in bad spirits, felt constrained. In his present frame of mind, he felt uncomfortable and ill at ease in a restaurant with its private rooms…continuous bustle and running about, these bronzes, mirrors, gaslights…all this seemed an affront to him…
>
> "Me? Yes, I am. Besides, all this makes me feel uncomfortable," said Levin. "You can't imagine how strange all this seems to one who lives in the country."[824]

---

823 Jack Weatherford, *Savages and Civilization: Who Will Survive?* 1994, 289.
824 Leo Tolstoy, *Anna Karenina*, 51.

Oblonsky and Levin continued their conversation with observations about the city, and the length of fingernails and of meals. After Levin had expressed his skepticism of "all this" in the restaurant, Oblonsky made the dubious remark:

> But that's the whole aim of civilization: to make everything a source of enjoyment.

Country boy Levin immediately rejoined, concluding:

> Well, if that is so, I'd rather be a savage.

Levin made further remarks about spending money, the means whereby we can make "everything a source of enjoyment," an alluring but awfully false promise.

Another of Weatherford's observations fits well into what I have been trying to say. "We cannot go backwards in history and change one hour or one moment, but we do have the power to change the present and thus alter the future."[825] Many others have been saying that unless we decide to make conscious changes in the present, there will be no future for ourselves, for our children, or for our place on Earth.

Scott Peck felt that we stand on "the brink of self-annihilation."[826] We will push or lead along our fellow creatures into oblivion, as we have already been doing. However, if we can learn to see our individual selves as part of the problem *and* the solution, we can each be more active in charting a healthier way ahead.

The Twelve Step pioneers hoped for worldwide success and their movement is partway along to achieving that goal. The desire to carry the message around the world is nowhere better expressed than in Bill W.'s 1959 article, "AA Communication Can Cross All Barriers."[827] After

---

825 Jack Weatherford, *Savages and Civilization*, 290.
826 M. Scott Peck, *The Different Drum*, 17.
827 *The Language of the Heart*, 292–96.

On Evolution

overseas travel with his wife, Lois W., Bill W. wrote that the Twelve Step way of life was already crossing the barriers of culture, religion, language, and race, and that it would perhaps help to heal the wounds of war.[828]

Bill W. noted that skepticism over the usefulness of the spiritual Twelve Steps could result from, quoting Herbert Spencer, "contempt prior to investigation."[829] There is too, some opposition to what is sometimes seen as a patriarchal A.A.[830] However, A.A., though founded mainly by men, is not patriarchal - nor perfect. It is, as one will find at a Twelve Step meeting that one likes, regularly attended, egalitarian and non-authoritarian. If one meeting does not live up to expectations, try another. Shop around to find a meeting that fits for you.[831] You are sure to find a group that you like. If not, start a new one.

The disenchanted can attend A.A. meetings for women, for people of colour, for First Nations people, or for the GLBTQ (Gay, Lesbian, Bisexual, Transgendered, and Queer) community. Skeptics can look to the critical role of Lois W. and Anne S., who stood beside Bill W. and Dr. Bob, keeping them alive so that they were able to create A.A. Those opposed to Twelve Step programs might still be able to recognize the potential of HPMA for good, by attending speaker meetings where there is no pressure to conform, or by reading Twelve Step Conference Approved literature and "outside" literature.[832]

Citing the divisiveness of world religions, Joseph Campbell said, "some mythology of a broader, deeper kind than anything envisioned

---

828 Ibid, 294–96.

829 *Alcoholics Anonymous* (The Big Book), 2001, 568

830 See Charlotte Kasl, *Many Roads, One Journey*, 1992; or Marianne Gilliam, *How Alcoholics Anonymous Failed Me: My Personal Journey through Self-Empowerment*, (New York: Eagle Brook, 1998).

831 Spiritual shopping is written about in Al-Anon. (*Having Had a Spiritual Awakening...*, 132).

832 Coming to mind is Nowinski's science based approach to the evaluation of A.A.'s effectiveness. (Joseph Nowinski, *If You Work It, It Works! The Science Behind 12 Step Recovery*, 2015). Similarly, Nancy Ellen Abrams has published *A God That Could Be Real: Spirituality, Science, and the Future of our Planet* (Boston: Beacon Press, 2015).

anywhere in the past is now required."[833] Arnold Toynbee suggested the establishment of a *modus vivendi*, a way of living by which the world's religions could live together in peace.[834] Such a way of living has always been available to us in the Higher Powered Mutual Aid of ancient First Nations societies, and now in its recent resurrection by the Twelve Step pioneers. HPMA offers a spiritual way of living that can accommodate any who pass through its broad doors.

On the one-hundredth anniversary of the Nobel Prize, one hundred Nobel laureates said, appropriately:

> The only hope for the future lies in cooperative international action, legitimized by democracy...To survive in the world we have transformed, we must learn to think in a new way. As never before, the future of each depends on the good of all.[835]

Learning to think and act in a better way has long been offered by First Nations and Twelve Step Higher Powered Mutual Aid.

Gregory Bateson had much praise for Bill W. He thought that A.A.'s cofounders had "extraordinary vision."[836] He saw Bill W. as "a very clever man, very clever,"[837] and is quoted as calling him a "master psychologist."[838] Aldous Huxley, author of *Brave New World*, called Bill W. "the greatest social architect of the [20th] century."[839] In 2017,

---

833 Joseph Campbell, *The Masks of God: Primitive Mythology*, 18.

834 Miguel Serrano, *C.G. Jung and Hermann Hesse: A Record of Two Friendships*, (London, UK: Routledge & Kegan Paul, 1966), 71.

835 "Our best point the way." *Toronto Globe and Mail* (December 7, 2001), A19.

836 Gregory Bateson, *Steps to an Ecology of Mind: Collected Essays in Anthropology, Psychiatry, Evolution, and Epistemology*, (San Francisco: Chandler Publishing Co., 1972), 310.

837 Gregory Bateson et al, *Angels Fear - Toward an Epistemology of the Sacred* (New York: Macmillan, 1987), 128.

838 Peter Harries-Jones, *A Recursive Vision: Ecological Understanding and Gregory Bateson*, (Toronto: University of Toronto Press, 1995), 40.

839 *'PASS IT ON': The story of Bill Wilson*, 368.

On Evolution

Russell Brand had much praise for the Twelve Steps, describing Bill W. as a prophet.[840]

Bill W. was to be presented with six honorary degrees during his lifetime, and the American Public Health Association's Lasker Award, but he humbly declined these honours for himself.[841] The genius was also considered for the Nobel Prize.[842] If it had been offered to him, deservedly so, Bill W. would no doubt have politely declined, and suggest that the award be given to Alcoholics Anonymous, the first Twelve Step society for whose success he worked so tirelessly.

---

840 Russell Brand, *Recovery: Freedom from Our Addictions*, (London, UK: Pan Macmillan, 2017), 8.
841 *'PASS IT ON': The story of Bill Wilson*, 311-13, 350.
842 Matthew Raphael, *Bill W. and Mr. Wilson*, 2000, 177; Susan Cheever, *My Name Is Bill*, 2004, 191.

Appendix A
Kropotkin's Wider Influence

There is evidence that Bill W. employed more important words and ideas from the "gentle Russian prince," Peter Kropotkin, beyond those already cited. He emphasised that the principles of A.A. recovery "...are borrowed, and so are most of our structural ideas."[843] He also said that A.A. had derived its principles - *"every one of them"*[844] - from many sources, mostly unnamed. After the Cold War with the U.S.S.R, he may have wanted us to do the research on the subject, in safety. At one point, again, Bill W. stated that he hoped that pioneering in A.A. would never end. [845]

A reading of Peter Kropotkin's *Mutual Aid: A Factor of Evolution* uncovers key words and phrases that Bill W. may have adopted for use in his writings about A.A., in *The Language of the Heart* and elsewhere. This seems particularly so regarding the development of the Twelve Traditions, ratified by A.A. in 1951. Eminent A.A. researcher and writer, Ernest Kurtz, recommended a "close" reading of *Mutual Aid*.[846]

When writing of A.A.'s benign anarchy in *Alcoholics Anonymous Comes of Age*,[847] Bill W. would not have failed to see the optimism in Kropotkin's detailed but readable, *Mutual Aid*, and other books, defining an ecological morality. However, First Nations and Twelve Step societies have a healthier spiritual and ecological planetary morality. This is where the Twelve Step pioneers parted company with Kropotkin and took up with the beliefs of William James and Carl Jung.

The three following sections contain words and phrases used in Bill W.'s writings that are the same as, or very similar to, language used

---

843 *Alcoholics Anonymous Comes of Age*, 1957, 224.
844 Ibid, 231.
845 Ibid, 80.
846 Ernest Kurtz, *Not God: A History of Alcoholics Anonymous*, 1979, 330, note 4.
847 *Alcoholics Anonymous Comes of Age*, 224–25.

by Peter Kropotkin in *Mutual Aid*. Use of this language is presented chronologically in Bill W.'s writings.

Mutual Aid

Bill W.'s first published use of the term "mutual aid" may have been in *The Language of the Heart* article of September 1945, "'Rules' Dangerous but Unity Vital."[848] This is the same article in which he described alcoholics as "true anarchists at heart." Here, Bill W. used "mutual aid" twice while writing of the failed Washingtonian Society, describing their use of mutual aid as a positive force, though weakened by insufficient emphasis on personal change and spirituality.

With his independent Vermont Yankee and New York stock market background, Bill W. may not have liked the word mutual, and for a good while after the 1945 article, his *The Language of the Heart* writings touched on other aspects of the Traditions. More importantly, in the atmosphere of the Cold War with the U.S.S.R., more use of the phrase mutual aid - the short title of Kropotkin's book - might not have been prudent.

Bill W.: My First 40 Years (2000) uses powerful imagery to explain how Ebby T. had been able to help him in New York in 1934, simply because Ebby was an alcoholic, too.[849] Bill W. did not use the term mutual aid to describe this experience, and, unfortunately, the 1954 recording - intended partly "to set the record somewhere near straight"[850] - ended without mention of his Twelfth Step call to Dr. Bob, carrying the mutual aid message to him in Akron, Ohio in the spring of 1935.[851]

In *Alcoholics Anonymous Comes of Age*, Bill W. described his early experiences of one alcoholic talking to another, drawing a distinction between the conversations he had with Ebby T. and later with Dr. Bob. Here, Bill W. summarized his meeting with Ebby by saying, "[in] the

---

848 *The Language of the Heart*, 6–9.
849 *Bill W.: My First 40 Years*, 153–54.
850 Ibid, 2.
851 Akron, Ohio is the birthplace of Alcoholics Anonymous.

On Evolution

kinship of common suffering, one alcoholic had been talking to another."[852] But with Dr. Bob, Bill W.'s experience is described in this manner:

> You see, our talk was a completely *mutual* thing. I had quit preaching. I knew that I needed this alcoholic as much as he needed me. *This was it*. And this mutual give-and-take is at the very heart of all of A.A.'s Twelfth Step work today. This was how to carry the message. The final missing link was located right there in my first talk with Dr. Bob.[853]

In the article of March 1960, "After Twenty-Five Years," Bill W. said that, through his conversations with Ebby T., he had partly realized the tremendous positive effect of one alcoholic talking to another.[854] But in the same paragraph, Bill W. added that the importance of one alcoholic talking to another in "mutuality" became fully clear to him only when he, now sober and spiritual, talked to drinking but spiritual Dr. Bob. Their spirituality at the time was evident with their participation in the Christian Oxford Group.

In the July 1960 article, "The Language of the Heart," Bill W. said that he and Dr. Bob shared the insight that real communication must be rooted in "mutual need."[855] Dr. Bob himself immortalized the message of mutual aid that he'd received in a passage from "Doctor Bob's Nightmare," the only words italicized in his story in the Big Book.[856] Bill W. was making frequent use of mutual aid, mutuality, and mutual need by the late 1950s and early 1960s.

The early struggles against alcohol were so difficult that men, women, and even their children, worked together to help each other, with the guidance of Higher Powers. It is not often noted that the children of Anne

---

852 *Alcoholics Anonymous Comes of Age*, 59.
853 Ibid, 70.
854 *The Language of the Heart*, 298.
855 Ibid, 247.
856 *Alcoholics Anonymous* (The Big Book), 180.

S. and Dr. Bob attended, too.[857] A.A.'s biography, *Dr. Bob and the Good Oldtimers*, directly acknowledged that the work of the wives of alcoholics had been essential to the birth and growth of A.A.[858]

Lois W., said Hartigan, felt that "all wives who joined their husbands in working with alcoholics during AA's formative years saw themselves as AA members."[859] Along with their involvement in early A.A., the wives of mostly male alcoholics started mutual aid family groups as early as 1936 in Akron and around the same time in New York.[860] In 1951, the Al-Anon Family Groups, for the family and friends of alcoholics, formally became the second Twelve Step fellowship or society.

Al-Anon's Twelve Steps differed, as earlier noted, from those of A.A. by only one word, "others" replacing "alcoholics" in the Twelfth Step. But Al-Anon's Twelve Traditions were extensively re-worded, first approved and ratified by A.A. in 1955 and by Al-Anon in 1961.[861] Further, Lois W. said that, in the careful work of adapting the Al-Anon Traditions from those of A.A., "Bill helped me a lot with this."[862] Indeed, it is clear that Al-Anon wanted A.A.'s approval of their Traditions,[863] and the four mentions of A.A. in the Al-Anon Traditions are one result.

A.A. was thus likely to have been helpful in the inclusion of Kropotkin's term "mutual aid" in Al-Anon's Third Tradition, as follows: "The relatives of alcoholics, when gathered together for mutual aid, may call themselves an Al-Anon Family Group, provided that, as a group, they

---

857 Bob Smith et al, *Children of the Healer: The Story of Dr. Bob's Kids* (Park Ridge, IL: Parkside Publishing Corporation), 1992.

858 *Dr. Bob and the Good Oldtimers* (New York: A.A. World Services, Inc., 1980), 235.

859 Ibid, 235; Francis Hartigan, *Bill W.: A Biography of Alcoholics Anonymous Cofounder Bill Wilson*, 71.

860 Bob Smith et al, *Children of the Healer*, 132; Robert Thomsen, *Bill W.* (Center City, MN: Hazelden, 1975), 273; *First Steps: Al-Anon . . . 35 Years of Beginnings*, 62.

861 *Lois Remembers*, 178; *First Steps: Al-Anon . . . 35 Years of Beginnings*, 85; *How Al-Anon Works for Families and Friends of Alcoholics*, 128.

862 *Lois Remembers*, Virginia Beach, VA: AFG Headquarters, Inc., 177.

863 Ibid, 178; *First Steps: Al-Anon...35 Years of Beginnings*, 80.

have no other affiliation. The only requirement for membership is that there be a problem of alcoholism in a relative or friend."

Implicit use of mutual aid had always been made by A.A.: alcoholics helping other alcoholics and themselves. Al-Anon's work was to be carried out with its explicit use. The crafting of Al-Anon's Third Tradition showed that the first two Twelve Step societies were prepared, by the 1950s, to more openly acknowledge and use Peter Kropotkin's important phrase, mutual aid, from *Mutual Aid: A Factor of Evolution*.

Self-Sacrifice for the Common Welfare

From his reading and research, Bill W. might have seen that Peter Kropotkin had refused a prestigious scientific position in 1871, in order to engage in the struggle for the common welfare in Russia.[864] Bill W. sacrificed himself for the common good of a New York group led by Lois and him, refusing an offer of employment in 1936 when he and Lois W. were dreadfully poor.[865] If nothing else, this showed that the two men were of like minds regarding the importance of making personal sacrifices for the common welfare.

Early on in his *Mutual Aid*, Kropotkin made the following statement:

> However terrible the wars . . . and whatever the atrocities committed at war-time, mutual aid within the community, self-devotion grown into a habit, and very often self-sacrifice for the common welfare, are the rule.[866]

One might easily think that Kropotkin was speaking of human beings here, but he was actually describing the behaviour of Brazilian ants. According to Kropotkin, "self-devotion grown into a habit" meant a focus on the individual ant's own health, purpose, and individuality within the

---

864 Peter Kropotkin, *Memoirs of a Revolutionist*, 234–41.
865 *'PASS IT ON': The story of Bill Wilson*, 175–77.
866 *Mutual Aid: A Factor of Evolution*, 14.

ant community. Individuality "in relation" is well preserved in societies of Brazilian ants. Morris says that "Kropotkin was . . . concerned to stress - even for insects - individual initiative and conscious agency."[867]

In every chapter of *Mutual Aid*, Kropotkin referred to examples of self-sacrifice for the common welfare. He observes the traditions of self-sacrifice among humans of common purpose, in tribes and clans,[868] and in fishing and mining villages.[869] He noted how the tradition of self-sacrifice for the common welfare sometimes strengthens during stressful times,[870] the recent case in point being the rise of Alcoholics Anonymous during the Great Depression of the 1930s.

But such traditions of self-sacrifice for the common welfare, Kropotkin noted, break down where ethical bonds of common interest and purpose are weakened, for instance, sometimes in cities.[871] Kropotkin cited an example of a contrast between the wealthy and the poor of his time, saying:

> Some training - good or bad, let them decide it for themselves - is required in a lady of the richer classes to render her able to pass by a shivering and hungry child in the street without noticing it. But the mothers of the poorer classes have not that training. They cannot stand the sight of a hungry child; they must feed it, and so they do.[872]

Kropotkin's descriptions of self-sacrifice for the common welfare may have been a great influence on Bill W. Early on, he used the phrase self-sacrifice in only one place that I have found, probably before encountering

---

867 Brian Morris, *Kropotkin: The Politics of Community* (Amherst, NY: Humanity Books, 2004), 182.
868 *Mutual Aid*, 110–13.
869 Ibid, 275–77.
870 Ibid, 96.
871 Ibid, 277.
872 Ibid, 285-86. Kropotkin does not say how *he* dealt with these children.

On Evolution

Kropotkin.[873] Kropotkin's extensive use of the phrase may well have confirmed its value to Bill W. Again, if nothing else, this too indicates that the two men were of like minds.

"Self-sacrifice" and "common welfare" were ideal words by which A.A. might deal with self-admitted, over-inflated egos, to allow A.A. to flourish as a society of common purpose; "ego-deflation at depth" is an important phrase in A.A. Equally, this enabled Bill W. to explain that the common welfare comes first, but that individuality is a close second.[874] He saw, too, that self-sacrifice for the common welfare broadened the purposes of anonymity to include humility.[875] Emphasis on humility would become of great benefit for the common good of Twelve Step HPMA groups, with no human authority and where leaders are only "trusted servants."

The phrases "self-sacrifice" and "common welfare" became important in describing the recovery process in A.A. There are many references to self-sacrifice, the common welfare (or common good), and anonymity in the literature of the first two Twelve Step fellowships. The first of these references may be from the mid 1940s in *The Language of the Heart* articles by Bill W., in which he also mentioned anarchy and mutual aid.

The April 1946 article, "Twelve Suggested Points for AA Tradition," explains the start of Bill W.'s attempt to deal with the relationships between A.A. members and A.A. groups, between the groups and A.A. as a whole, and between A.A. and the fragile global society of post-World War II.[876] Bill W. set down the "Twelve Points" that, with modifications, would be ratified in 1950 as the short form of A.A.'s Twelve Traditions. "Our common welfare comes first" is contained in the First of the Twelve

---

873 *Alcoholics Anonymous* (The Big Book), 93.
874 *The Language of the Heart*, 32.
875 *Twelve Steps and Twelve Traditions*, 184–87.
876 *The Language of the Heart*, 21.

Points.[877] The words "our common welfare is paramount" are also used in the Fourth Point.[878]

Looking back to this period, Bill W. called the development of the Twelve Traditions a daring undertaking, well knowing the danger of anything resembling rules among alcoholics.[879] As in the Twelve Points of *The Language of the Heart*, the Twelve Traditions (the long form) of *Twelve Steps and Twelve Traditions* (1953) were written to ensure that the common welfare comes first.[880] Tradition Four (of the long form), as in Point Four, says that the common welfare is supreme, describing how individual and group autonomy are circumscribed by the common welfare.[881]

The April 1946 article was followed by another, the July 1946 piece cited earlier, "The Individual in Relation to AA as a Group," possibly a response to readers' fears of a loss of individuality. Here Bill W. repeated the first Point exactly as it had been written earlier, in April. But now he was more forceful, saying it was characteristic of all societies that individuals must sometimes put the "common welfare"[882] ahead of their own, and that without this, no society could continue to exist. Kropotkin had said this in *Mutual Aid*.[883]

A subsequent article, dated November 1948, was called "Tradition Twelve." Here, Bill W. wrote that the individual or group needed to "give up" something for the good of A.A.[884] Bill W. may have given up something here too: the "Twelve Points" and the Twelve Traditions (the long form) say that the common welfare comes first, but A.A.'s Traditions (the short form) reads that it *should* come first.

---

877 Ibid, 22.
878 Ibid
879 Ibid, 302.
880 *Twelve Steps and Twelve Traditions*, 189.
881 Ibid
882 *The Language of the Heart*, 32.
883 *Mutual Aid: A Factor of Evolution*, 41–42.
884 *The Language of the Heart*, 93.

On Evolution

A.A.'s *Twelve Steps and Twelve Traditions* (1953) offers detailed guidance on the importance of individuality within the common welfare. That the common welfare should come first is emphasized especially between pages 129 and 131. Later, Bill W. linked self-sacrifice with anonymity, saying that the Twelve Traditions asked members, "repeatedly," to sacrifice personal needs for the general welfare.[885] In *Alcoholics Anonymous Comes of Age* (1957), the same words and sentiment appear again,[886] emphasizing the importance that Bill W. attached to them.

Self-sacrifice for the common welfare received its strongest treatment from Bill W. in the article of January 1955, "Why Alcoholics Anonymous is Anonymous." He wrote that members had to make "sacrifices for the common welfare" of A.A.[887] Part of this piece was repeated almost word for word in *Alcoholics Anonymous Comes of Age*, as below. They are some of Bill W.'s finest words, a call to arms against the self-fulfilling prophecy of the Darwinian war of the "struggle for existence" in the macrocosm of the larger world and in the microcosm of A.A.; Bill W. wrote:

> As never before, the struggle for power, importance, and wealth is tearing civilization apart - man against man, family against family, group against group, nation against nation.
>
> Nearly all those engaged in this fierce competition declare that their aim is peace and justice for themselves, their neighbors, and their nations. "Give us power," they say, "and we shall have justice; give us fame and we shall set a great example; give us money and we shall be comfortable and happy." People throughout the world deeply believe such things and act accordingly. On this appalling dry bender, society seems to be staggering down a dead-end road. The stop sign is clearly marked. It says "Disaster."

---

885 *Twelve Steps and Twelve Traditions*, 184.
886 *Alcoholics Anonymous Comes of Age*, 132.
887 *The Language of the Heart*, 210.

What has this got to do with anonymity, and Alcoholics Anonymous?

We of AA ought to know. Nearly every one of us has traversed this identical dead-end path. Powered by alcohol and self-justification, many of us have pursued the phantoms of self-importance and money right up to the disaster stop sign. Then came AA.

We faced about and found ourselves on a new highroad where the direction signs said never a word about power, fame, or wealth. The new signs read, "This way to sanity and serenity. The price is self-sacrifice."[888]

With the ratification of A.A.'s Twelve Traditions (the short form) in 1950, Bill W. extensively described the close association of the Traditions with self-sacrifice for the common welfare of Alcoholics Anonymous.[889]

The experience of the groups, and the pioneers' own knowledge and research, led A.A. members to see that anonymity had an importance rooted in more than shame and fear. The sacrificing of the ego for the common welfare enlarged the rationale for anonymity to include humility.[890]

Bill W. became explicit about this in *Alcoholics Anonymous Comes of Age*, saying that the pioneers had come to see that the effectiveness, unity, welfare, and survival of A.A. depended on members' willingness to make "sacrifices for the group, sacrifices for the common welfare."[891] One A.A. member wrote about the self-sacrifice of anonymity as a check on the individual ego.[892] Anonymity and humility began to serve the individual good and the common good, together, in Alcoholics Anonymous.

---

888 *Alcoholics Anonymous Comes of Age*, 286.
889 Ibid, 132, 286–88; *Twelve Steps and Twelve Traditions*, 129–31, 184.
890 *Twelve Steps and Twelve Traditions*, 187.
891 *Alcoholics Anonymous Comes of Age*, 287.
892 Alcoholics Anonymous, "The A.A. Group: Where It All Begins," pamphlet, 2005, 6.

Society

Alcoholics Anonymous was a "nameless bunch of drunks"[893] in the late 1930s, but it graduated to more respectable descriptions such as "fellowship" and later "society," the latter a word used extensively by Kropotkin in *Mutual Aid*. "Bill's Story" in *Alcoholics Anonymous* (The Big Book) has not changed since 1939, and in it, Bill W. describes A.A. as a fellowship; there is no mention of A.A. as a society. His foreword to the 1939 edition also calls A.A. a fellowship once, and, while he also used the word "societies" in the foreword, it was not to describe A.A.

Bill W. may have encountered Kropotkin's writings in the early to mid-1940s, and he began to use the word society to describe A.A. His July 1946 article, "The Individual in Relation to AA as a Group," may contain its first mention. The opening sentence says that A.A. might be "a new form of human society."[894] Society is used in this article six times, while fellowship is not used at all.

In the October 1947 article, "Why Can't We Join AA, Too?" Bill W. said that A.A. might be a "new kind of human society."[895] *The Language of the Heart* has many references to A.A. as a society,[896] as well as a fellowship, and from the late 1940s on, the two are almost interchangeable in Bill W.'s Grapevine writings.

The word society was used to describe A.A. in the foreword to the 1955 edition of the Big Book; it is used five times here, while fellowship is used twice. Society is used once more in the 1955 book, in Appendix I, "The A.A. Tradition";[897] this is also the case in newer editions. A.A.'s Twelve Traditions, both the short and long form, were not ratified until after 1939; thus, they were not included in the 1939 Big Book.

Since Bill W.'s passing in 1971, newer editions of the Big Book do not give the word society quite the same emphasis. The 1976 preface

---

893 *'PASS IT ON': The story of Bill Wilson*, 307.
894 *The Language of the Heart*, 32.
895 Ibid, 108.
896 Ibid, e.g., 36, 78, 89, 160, 200, 223, 287, 360.
897 *Alcoholics Anonymous* (The Big Book), 563.

employs society twice and fellowship once, while the 1976 foreword uses fellowship once but not society. The 2001 book preface uses society twice to describe A.A., and fellowship once, while the 2001 foreword uses fellowship twice and society not at all.

In the same paragraph where he refers to the Russian prince in *Alcoholics Anonymous Comes of Age*, Bill W. described A.A. as a "society" of alcoholics who "voluntarily associate themselves in a common interest."[898] This is an actual dictionary definition of society, and a means whereby Bill W. was able to more precisely distance A.A. from government. "Fellowship" was used first to define A.A. as a "body of associates," but it lacks the common interest and specific purpose of "society."

Bill W. might have found the word society anywhere, for example, through his study of the Washingtonian Society. But Kropotkin's extensive use of the word in *Mutual Aid* was from a scientific and evolutionary perspective. This would no doubt have appealed to Bill W. in his fashioning of words to describe an evolution of A.A. that was firmly rooted in a voluntary common interest.

Other Possible Influences from Kropotkin

More of Kropotkin's words are repeated below for their value in understanding the critical importance of mutual aid; Bill W. may have seen this, too. As earlier mentioned, Kropotkin wrote:

> In short, neither the crushing powers of the centralized State nor the teachings of mutual hatred and pitiless struggle which came, adorned with the attributes of science, from obliging philosophers and sociologists, could weed out the feeling of human solidarity, deeply lodged in men's understanding and heart, because it has been nurtured by all our preceding evolution. What was the outcome of evolution since its earliest stages cannot be overpowered by

---

898 *Alcoholics Anonymous Comes of Age*, 224–25.

one of the aspects of that same evolution. And the need of mutual aid . . . re-asserts itself again, even in our modern society, and claims its rights to be, as it always has been, the chief leader towards further progress.[899]

The Russian prince's optimistic vision for the future may have increased Bill W.'s hope for a better world through Twelve Step Higher Powered Mutual Aid. Bill W.'s powerful interpretation of "in *all* our affairs" in *Twelve Steps and Twelve Traditions*, his 1961 article "The Shape of Things to Come," and his belief that A.A. might actually change the world,[900] may well have been sympathetic responses to Kropotkin's ideas, rooted in Bill W.'s own inspired and well-researched beliefs.

Bill W. had clearly gone through a sea change in his thinking about the world, from his early competitive days in the classroom, on the sports field, in the stock market, through to his experience of World War I, to his spiritual awakening and sobriety. As noted, he echoed words used by Russian prince Kropotkin in *Mutual Aid* (1902). The following are quotations from *Alcoholics Anonymous Comes of Age* and *Mutual Aid*, each carrying similar wording, style, and sentiment:

> As never before, the struggle for power, importance, and wealth is tearing civilization apart - man against man, family against family, group against group, nation against nation.[901]
>
> The old bonds seem entirely to be broken. Stems [family branches] are seen to fight against stems, tribes against tribes,

---

899 *Mutual Aid*, 292.

900 A.A.'s biography of Bill W. says that he "believed [A.A.] would change the world." (*'PASS IT ON': The story of Bill Wilson*, 265). Bill W.'s wife, Lois W., said it was no exaggeration that her husband wanted to change the world for the better. (*First Steps: Al-Anon . . . 35 Years of Beginnings*, 1986, 156).

901 *Alcoholics Anonymous Comes of Age*, 286.

individuals against individuals [with] States always ready to wage war against each other . . .⁹⁰²

In *Alcoholics Anonymous Comes of Age*, Bill W. wrote that ideas from the Oxford Group, William James, and Dr. Silkworth helped the pioneers to formulate an initial six steps of A.A. by 1938.⁹⁰³ Bill W. later composed others for A.A.⁹⁰⁴ The new draft added up to twelve, and Bill W. wrote, "[T]his number seemed significant. Without any special rhyme or reason, I connected them with the twelve apostles."⁹⁰⁵ *'PASS IT ON'* says that he had simply expanded the six Oxford Group principles to the Twelve Steps.⁹⁰⁶

'PASS IT ON' also said that the number twelve was "symbolic,"⁹⁰⁷ but this is not expanded upon. The Western and Chinese zodiacs each have twelve signs, and the Chinese use a twelve-part system of healing.⁹⁰⁸ Kropotkin's *Mutual Aid* reads:

> The 'Twelve Articles' and similar professions of faith, which were circulated among the German and Swiss peasants and artisans, maintained . . . every one's right to interpret the Bible according to his own understanding . . .⁹⁰⁹

Bill W. might have liked the ring of "Twelve Articles" beside the Twelve Steps. Bill W., Dr. Bob, and their wives, Lois W. and Anne S. had engaged in spiritual practices of many kinds, including spiritualism, e.g.,

---

902 *Mutual Aid: A Factor of Evolution*, 115–16.
903 *Alcoholics Anonymous Comes of Age*, 160.
904 Ibid, 161.
905 Ibid
906 *'PASS IT ON': The story of Bill Wilson*, 197–98.
907 Ibid, 198.
908 Leon Hammer, *dragon rises, red bird flies: Psychology and Chinese Medicine* (Barrytown, NY: Station Hill Press, Inc., 1990), 37–38.
909 *Mutual Aid*, 225.

On Evolution

a "spooking circle,"[910] eager to try anything that might help alcoholics.[911] The final arrival at the Twelve Steps is still a bit of a mystery.

The Big Book records, as noted, that Ebby T. asked a skeptical Bill W. to consider choosing his own understanding of God.[912] William James's *The Varieties of Religious Experience*, published the same year as *Mutual Aid*, was read carefully by Bill W. and other Twelve Step pioneers.[913] James adopted a pluralistic approach to religion and to God,[914] while Kropotkin was opposed to religion. Yet, Kropotkin's use of ". . . every one's right to interpret the Bible according to his own understanding . . ." might also have been welcome words to the Twelve Step pioneers, reminding them of A.A.'s Steps Three and Eleven, where it says we can have the God of our own understanding.

Other important ideas contained in Kropotkin's *Mutual Aid* may have been influential in early A.A. It is probable that Bill W. and Dr. Bob - opposed to Franklin Delano Roosevelt's Depression-era big-spending initiatives - would have agreed with Kropotkin's analysis of the power of the centralized state to weaken self-regulation and self-responsibility in pursuit of the common welfare. In this context, Kropotkin wrote:

> The absorption of all social functions by the State . . . necessarily favored the development of an unbridled, narrow-minded individualism. In proportion as the obligations towards the State grew in numbers, the citizens were evidently relieved from their obligations towards each other.[915]

---

910 *'PASS IT ON': The story of Bill Wilson*, 278-80; Matthew Raphael, *Bill W. and Mr. Wilson*, 159.

911 Before LSD - lysergic acid diethylamide - became a banned substance, Bill W. thought it might be of some use, after trying it himself, and observing beneficial changes in others who'd also tried it, including his wife, Lois W. (*'PASS IT ON': The story of Bill Wilson*, 369-77).

912 *Alcoholics Anonymous* (The Big Book), 12.

913 *'PASS IT ON': The story of Bill Wilson*, 124.

914 William James, *The Varieties of Religious Experience* (New York: Mentor Books, 1958; originally published 1902), e.g., 264-66, 399, 413-14.

915 *Mutual Aid*, 227.

Words such as these may well have added strength to the practical foundations of the Twelve Step movement, where authority is vested in Higher Powers, not in human authority, in government, in business, or in religion.

Related to this are strong parallels between Kropotkin's disagreement with charity and the need for self-support in Twelve Step societies. Kropotkin said further:

> [W]hile early Christianity, like all other religions, was an appeal to the broadly human feelings of mutual aid and sympathy, the Christian Church has aided the State in wrecking all standing institutions of mutual aid and support which were anterior to it, or developed outside of it; and instead of the mutual aid which every savage considers as due to his kinsman, it has preached charity which bears a character of inspiration from above, and, accordingly, implies . . . superiority of the giver over the receiver.[916]

At a time when he was trying to lessen his influence in A.A., Bill W. says briefly in a letter, that dictatorships and hierarchical churches patronize people, taking away their self-responsibility, self-support, and the freedom to ". . . properly guide their own destinies [including] the opportunity of really growing up."[917] Bill W. was being exhausted by his followers, who looked up to him instead of seeing him as a "trusted servant."

Kropotkin's writing may well have confirmed what Bill W. and the pioneers had already experienced, talked of, heard, and read, proving that mutual aid is as natural, evident, and much more widespread than competition. The two men thought the same on some critical issues. The evidence indicates that Bill W. and the other A.A. founders used ideas from the Russian prince to move A.A. and the Twelve Step membership further toward self-regulation, self-respect, self-responsibility, self-sacrifice, and self-support.

---

916 Ibid, 283.
917 *'PASS IT ON': The story of Bill Wilson*, 373. "Growing up" in First Nations, A.A., and Al-Anon is about learning all-important self-responsibility.

On Evolution

While praising the twin pillars of religion and medicine for their support of A.A., Bill W. played down the role of many others, including prince Kropotkin. But the avoidance of controversial outside issues - such as Kropotkin's anarchism - was important for the survival of individual members and for A.A. itself.

As noted earlier, it might well have been disastrous for the Twelve Step founders to cite the work of the anarchist prince extensively during the Cold War. This was a measure of Bill W. and the pioneers at their most astute. But, in the end, Twelve Step Higher Powered Mutual Aid societies actually became quite different from the type of mutual aid groups that Peter Kropotkin envisioned, wrote about, and practiced, but which were unsuccessful, as Kropotkin himself admitted.

Appendix B
Darwin was not an ecologist

Charles De Paolo's research on Charles Darwin's ethnographic writings makes it clear that Darwin had unfortunate racist tendencies; this was described in Chapter Three. Darwin's attempt to provide a living "missing link" between apes and humans, led to the de-humanizing of peoples such as the Fuegians (Yamana) of Tierra del Fuego.

What else did this dehumanizing mean, employed not only by Darwin, but also by other Europeans who felt that Indigenous peoples would, or could, be exterminated in the white conquerors' campaigns for world domination? Noted earlier, before their first contacts with First Nations peoples, the Europeans were sick, living under the illusion, for one, that they were superior to others.

The Europeans were sick, also, with the illusion that they were not "in relation" anymore, that they were separate-from-Nature "individuals," as we have seen a myth on its own. They were among the first savages[918] to disconnect, *en masse*, from Nature, and, thus "freed," felt that they could legitimately vanquish the Earth with their purported better civilized societies and greater Gods.

The rise of alcoholism may well have had something to do with this. Rum, whisky, and other spirits began to quench European thirst before, during, and after the years of colonial expansionism,[919] perhaps, in a way, providing fuel for it. Excessive alcohol consumption may have had the unintended effect of separating Europeans from others even more, promoting a superiority complex among them, and violating the ecological principle of the "individual in relation" to Earth, Water, and Sky. They had broken away a part of themselves, their own lands, faced

---

918 In its meaning as "cruel," the Europeans were definitely savage,
919 Tom Standage, *A History of the World in 6 Glasses* (New York: Walker & Company, 2005), 4.

an increase in mental, moral, psychological, and physical illnesses, and committed the global murder of Indigenes in the name of "progress."

By chance, the sick Europeans met up, early on, with perhaps the *most* advanced peoples of the world, not the *least*, the First Nations peoples of the Americas. These were peoples who had migrated the farthest from Africa,[920] our original home, learning along the way how to "live with Nature" to the nth degree, developing, for example, an "exceptional knowledge of botanical psychopharmacology"[921] and related methodology. These peoples produced superior governments, too, giving benign anarchism a good name. One need not look much further than the Iroquois, as noted, who inspired U.S. and U.S.S.R. constitutionalists. Cavanagh et al wrote of an Iroquois submission to the United Nations re: the ownership of common lands and resources.[922]

There is among First Nations peoples an advanced unity between them and their ancestral lands, something missing from the cultures of almost everyone today, leaving us on the edge of a dangerous precipice. Mander et al cite examples of this in *Paradigm Wars*.[923] Winona La Duke, of the Anishnabeg tribe, contributed the evocatively titled article "The People Belong to the Land" to Mander's book.[924] Jeanette Armstrong, an Okanagan, wrote "Community: 'Sharing One Skin,'"[925] saying that "The soil, the water, the air, and all other life forms contributed to be our flesh.

---

[920] Marean is one of the latest to comment on our origins, possibly at southern coastal South Africa. Curtis Marean, "The Most Invasive Species of All," *Scientific American*, Vol. 313, No. 2, 33-39 (August 2015).

[921] Don Coyhis et al, *Alcohol Problems in Native America*, 2006, 9.

[922] John Cavanagh et al, eds., *Alternatives to Economic Globalization*, 2004, 129.

[923] Jerry Mander et al, eds., *Paradigm Wars: Indigenous Peoples' Resistance to Globalization*. International Forum on Globalization. (San Francisco: Sierra Club Books, 2006).

[924] Ibid, *Paradigm Wars*, 23–25.

[925] Ibid, 35–39.

On Evolution

We are our land/place."[926] That connection to land can apply in current farming society, too.[927]

Displaced from ancestral lands by white cruelty and exterminatory wars, First Nations people were forced to migrate to places that they did not know, in a fundamental sense. When Darwin found the Fuegians living in misery in inhospitable Tierra del Fuego, they may well have been recent immigrants, fleeing onslaught by the Spanish and Portuguese to the north. De Paolo notes that despite the long distances that separated them, the Fuegians were in fact related to the northern Patagonian Pehuenche.

The Fuegians were sick, too, mentally and physically, and, when FitzRoy and Darwin found them, were a prime example of what might be termed a "displacement illness." The Fuegians were miserable and showed it. However, later with Kropotkin at the turn of the century, Gusinde in the 1930s, and Bridges in the 1940s, the Fuegians had become more settled in what was to them initially foreign land in Tierra del Fuego.

The savage Europeans were missing an integral part of themselves, too, roots in the complex land, and were ill themselves. That displacement illness made them "out laws", notable for their efficient power to conquer others' land, and to kill. That way of life is what we have today, almost worldwide, a killer culture[928] that is never sated, "soulless" and insane, eliminating all that gets in its way. Wright describes us as "serial killers,"[929] at "an all-you-can-kill barbeque."[930]

---

926 Jerry Mander et al, eds., 37. A related observation by Daschuk is that the rise of transportation by horse among some First Nations "…marked the end of an ecological relationship between [them] and their [land] that was thousands of years old. (James Daschuk, *Clearing the Plains*, [Regina, SK: University of Regina Press, 2013], 42). It seems that that relationship has been, or is being, re-cultivated.

927 A friend and retiring farmer hoped for only one thing: that the next farmer would "take care of the land." Said one woman of her husband: "This was part of him - these fields - this farm." (Gwen Morris, *Gwen: Memories and Sketches*, Editor John Morris, [Alexandria, ON: The Printing House, 2014], 47).

928 This view is shared by Charlotte Kasl in her *Many Roads, One Journey*, 1992, 54.

929 Ronald Wright, *a Short History of Progress*, 2004, 43, 63.

930 Ibid, 39.

Not all is lost, yet. Why was Darwin, who explained the theory of evolution, now fact, unable to understand the serious implications of the displacement of Indigenous peoples by the Europeans? The answer is simply this: Darwin was not an ecologist,[931] "not yet an accomplished naturalist," as De Paolo says,[932] before his round the world trip. A *National Geographic* special on Darwin, aired September 8, 2013 on Television Ontario (TVO), referred to him as an "amateur naturalist" when he set out on the *Beagle*.[933] Says Korey, Darwin was a "neophyte scientist."[934] Darwin admitted this, saying that his journey "…was the first real training or education of my mind."[935] Darwin did not know of the central intimate-knowledge-of-the-sacred-land aspect of First Nations cultures. At best, he underestimated the effects of the displacement of First Nations people from ancestral land, land of which they considered themselves a part.[936] Said another of Mander's writers, Leslie Gray, "It can sound very strange to a [First Nations] person to hear [non-First Nations people] refer to the Middle East as the Holy Land. *This* is the Holy Land."[937]

Another case of Darwin's limitations is the much-ballyhooed notion of the supposedly different species of birds that he collected on various islands in the Galapagos archipelago. It was thought that these particular birds varied according to the size and shape of their beaks, and that they

---

931 Ernst Haeckel first used the term "ecology" in 1858, a year before Darwin published the first edition of *The Origin of Species*. There were five more editions, the last in 1872, before Darwin's death in 1882.

932 Charles De Paolo, *the Ethnography of Charles Darwin*, 61.

933 Darwin revealed this amateurism partway through his journey, saying that a particular Galapagos lizard was "ugly," "stupid," and "lazy." (Robert Jastrow, General Editor, *The Essential Darwin*, [Toronto: Little, Brown & Company (Canada) Ltd., 1984], 41).

934 Kenneth Korey in Robert Jastrow, General Editor, *The Essential Darwin*, 1984, 1.

935 Charles Darwin in Robert Jastrow, General Editor, *The Essential Darwin*, 1984, 18.

936 In "Indigenous Ecological Knowledge," Posey states, "[A]bove all, healthy ecosystems are critical to healthy societies and individuals, because humanity and nature are one." (Jerry Mander et al, eds., *Paradigm Wars*, 2006), 32.

937 Leslie Gray, "The Whole Planet Is the Holy Land," (Jerry Mander et al, eds., 2006, 28-29).

were therefore different species. However, the variability of the birds' beaks *within* islands (intra-island) may have proven equal to the variability of the birds' beaks *between* islands (inter-island). If this were so, there would not have been different species of these particular birds in the Galapagos.

In fact, the birds could easily have flown from island to island; the five largest are only 2 to 20 miles apart. While sailing on the *Beagle*, Darwin found "countless" butterflies 10 miles off the Patagonian coast, and a grasshopper 370 miles from Africa, both of which had apparently taken flight rather than being blown off shore.[938] Of course, this does not mean that the theory - fact - of natural selection is wrong. Darwin or others simply used a poor example to illustrate it.

As earlier noted, Darwin wrote in *The Origin of Species* (1859), that "No one ought to feel surprise at much remaining as yet unexplained in regard to the origin of species . . . if he makes due allowance for our profound ignorance in regard to the mutual relations of all the beings which live around us…relations [which] are of the highest importance."[939] He repeated this later, referring to "our ignorance on the mutual relations of all organic beings . . ."[940], namely, for example, ecological considerations.

In *The Origin of Species*, Darwin states, "…it deserves especial notice that the more important objections [to his theory of evolution] relate to questions on which we are confessedly ignorant; nor do we know how ignorant we are."[941] Furthermore, Darwin admitted in *The Descent of Man*, that many of his views were "highly speculative, and some no doubt will prove erroneous,"[942] taking himself down from the pedestal on which scores still seek to place him.

---

938 Charles Darwin, The *Voyage of the Beagle*, first published 1845, republished 1997, 159.
939 Charles Darwin, *The Origin of Species*, 1968, 1st published 1859, 68-69.
940 Ibid, 129.
941 Ibid, 440.
942 Charles Darwin, *The Descent of Man*, 1871, revised edition 1874, 601.

Appendix C
Nomad

Where does this leave us then, after recognizing the importance of "place" in our make up?

Is there something in our natures that is as important as alcoholism, or more so? Rene Dubos wrote about nomadism in 1959: "Prehistory and ancient history show that men have never been able to forget their nomadic past and to rest quietly in the corner of the earth they had made their own for a while."[943] We have always been wanderers, and we might have to honour this part of our natures.

The *healthy* nomadism of the small, *slowly moving* bands of mother-centric peoples,[944] originating in Africa,[945] has surrendered, in the end, to *unhealthy* nomadism that promotes the rapid separation from land, a

---

943 Rene Dubos, *Mirage of Health: Utopias, Progress, and Biological Change*, (New Brunswick, Rutgers University Press, 1959), 263.

944 "The image of brutal, domineering cave men is false. They were hunters and gatherers, wandering in small, peaceful groups formed around mothers, dependent on each other for survival." (*Goddess Remembered*, video, National Film Board of Canada, 1989). Wright mentions the "family bands" of early hominids (Ronald Wright, *A Short History of Progress*, 2004, 17).

945 This fits with genetic research showing that the first human was in fact an African woman. (R. L. Cann, M. Stoneking, and A. C. Wilson, "Mitochondrial DNA and Human Evolution." *Nature* v. 325 (6099): 1987, 31-36). There began a furor of controversy over this, one example being the work of Stringer; he felt that, instead of one woman, "there must have been thousands of women alive at that time." (Christopher Stringer et al, *African Exodus: The Origins of Modern Humanity*. [New York: Henry Holt and Company, 1996], 123). The controversy was further addressed by Cann in her "Mothers, Labels, and Misogyny." (Lori D. Hager, ed., *Women in Evolution*. [London: Routledge, 1997], 76-90). But with the evidence piling up, Stringer changed his mind, saying that the conclusions drawn by Cann et al, "…were essentially correct…" and that this "…female gave rise to all…humans today." (Christopher Stringer, *The Origin of Our Species*, [UK: Allan Lane, 2011], 173).

rootlessness probably originating in Europe.[946] This latter nomadism, promoting the life of the so-called "individual," is alive and well, with people now spread out to overpopulate the entire globe.

True nomadism, practiced in the developed world especially by tourists, and promoted everywhere else, does not recognize boundaries. The healthy *near*-nomadism that recognizes others' boundaries is practiced mainly by a decreasing number of Indigenous peoples in the developing world.

The Euro-American-centric developed world is responsible for promoting the seductive so-called "easy life" that all aspire to, around the world. Bending the rules of nature has become second nature. Such rule bending cheats Mother Nature, and tries to force Her to play the game of destruction that we, Her children, have insisted upon for the past 3000 years, and particularly the last 500. We have no problem with the profligate rape of Her skin, through boring for oil, mining, and "fracking" for instance, leading to the creation of products that pollute the air, land, and water, and to the creation of the "throw away" global society that we have today.

Exponential human population growth, accompanied by the devastating loss of other species' numbers, is, on the face of it, a greater problem than unhealthy nomadism. However, I suggest that the huge numbers of people we have today may instead be related to the loss of the natural world in the psyche of the unhealthy nomad in its global, not local, vision of itself. We forget that we are, and have always been, intimately related to land, that born from dust, we shall return to dust.[947] We are as a nation of wandering souls, tourists, separate from ancient lands that our forebears used to call home.

That "individual" nomad is us, primarily the powerfully influential, wealthy elites of the world, bereft of a real and necessary relationship to

---

946 We cannot here explore the history of Europe, replete with invasions from powers such as the Romans, Genghis Khan, and their followers.

947 The Bible says: "...the Lord formed man of the dust of the ground." (Genesis 2:7), and "...dust thou art, and unto dust shalt thou return." Genesis 3:19

On Evolution

the land, always moving, often and erratically, around the planet, insisting that everyone play the rules of transaction as they define them. That nomad never existed among the Europeans before they were sick enough to sever their own relationships with land. This resulted in the displacement of most Indigenous peoples from their land, around the globe. Despite the worldwide devastation caused by foreign cruelty and exterminatory invasions, Indigenous people are, thank God, making a timely resurgence. They have much to teach.

Mander et al feel that, "[D]espite the assaults upon Indigenous peoples, the technologies of development and materialism that are arrayed against them, and a global economic system that has conspired in every possible way to undermine Indigenous rights, Indigenous peoples are now everywhere resistant, well organized, and optimistic about eventual success."[948] This seems overly optimistic.

According to a 2011 Amnesty International (Canada) report, "By any measure, the failure to protect and uphold the rights of Indigenous people . . . remains [Canada's] most serious and longstanding human rights failure."[949] The 2014 Amnesty annual report zeroes in on the "Stolen Sisters" campaign, raising awareness about the missing and murdered Indigenous women of Canada.

Canada is one of the wealthiest nations in the world, but 2011 CBC Radio One broadcasts have reported that *116* Indigenous communities are under "boil-water" advisories,[950] and that Inuit men of a certain age are forty times more likely to commit suicide than other men. Despite Mander's optimism, there is much progress yet to be made in the treatment of Indigenous peoples, so that they can participate in and promote the recreation and re-blossoming of the planet.

---

948 Jerry Mander et al, *Paradigm Wars*, 2006, 223.
949 Amnesty International (Canada). *Getting Back on the "Rights Track": A Human Rights Agenda for Canada*. Ottawa, March 2011.
950 The number of boil water advisories remains at 100 (CBC-TV News, 10 p.m., October 5, 2017).

James G. Duncan

Today, we are ever ready to demonstrate our displacement from the land, living under the illusion of "individuality," sick with unhealthy nomadism. Most of us now live in cities, largely separate from Nature. However, a *true* "individual" can never be found, certainly not in this world. This individual is an actual illusion, something we do not want to know, or admit, so sick are we. We are always, actually and irrevocably, tied to the Earth and to others, in numerous ways, through the ecological commandment of the "individual in relation" to Creation. In all our wanderings from place to place, we carry the illusionary, sick, shadowy "individual" with us, not the "individual in relation."

The seemingly successful "individual" who thinks he is free, not attached to anyone or anywhere in particular, draws power from all he comes into contact with, but giving nothing in return, leaving spiritual, mental, and physical emptiness in his wake. The apparent health but spiritual impoverishment of this individual makes it seem that (s)he is leading an exciting, fulfilling life, possibly when (s)he mixes it with substance abuse.

"Ego-inflated," this individual, drunk with power drawn from elsewhere, sometimes thinks that he is God, always ready to violate the Earth's ecological law of the "individual in relation." Violating laws may be exciting to us, and may be part of our current adolescent makeup. It is past time for us to grow up and be our mature selves, "individuals in relation" to Creation, human and not God.

Does this mean that we will have to revert to older ways of living, each developing an intimate connection with the land? First Nations, farming, and gardening peoples have. Many will be able to do so locally, in their families and communities, but certainly not on a global scale. No, the important thing is to combat our unhealthy nomadism.

How can we do this? In two words: less nomadism. We will all have to make sacrifices, if trying to survive can be considered a sacrifice. To combat this nomadism, it will be necessary - with conscious purpose rooted in the unassailable fact of our existence as ecological, Earth-bound creatures - to re-cultivate our "individuality in relation" to others, and to Nature, *especially where we make our homes, gardens, and farms.* Carl Jung felt

On Evolution

that all should "have a plot of land so the instincts can come back to life again."[951] We can grow all our own food through advances in technology.

The boundaries of such areas, ideally, might be those of small watersheds, ecological as they are. Using such boundaries to define communities might be difficult, but it is far better than drawing straight lines that dissect Mother Earth. Small towns or urban enclaves in North America could sustain schools, stores, houses of worship etc., each a part of one's community where people can function on a face-to-face basis. The application of Higher Powered Mutual Aid in such places can bring us back, rooted to the Earth in healthier life,[952] and away from family, community, and world destruction.

In this vein, the globalization of trade in the world today is an important contributor to our unhealthy nomadism, even a menace to life as we know it. Here I am thinking of the physical aspect of globalization, actual travel to and/or the shipping of products to different countries, or to regions within large countries.[953] The costs of globalization are pollution, and weakened or destroyed families, communities, and countries.

Naively, we believe that unceasingly promoted globalization is a good thing, some claiming that it contributes to closer relations with others in the "global village."[954] The reality, however, is that we must instead increase our commitment to neighbourhood communities, and its products, while in large measure eschewing our nomadic lifestyles. People may grind their teeth at my ideas about travel,[955] but I see no other choice.

---

951 Meredith Sabini, ed., *The Earth has a Soul,* 154.

952 Said Carl Jung, too, "I am fully committed to the idea that human existence should be rooted in the earth." (Meredith Sabini, ed., *The Earth has a Soul,* 2008, 156).

953 Technology allows for the downloading of information, books, reports, etc., without paper.

954 Connections with other people around the world can come from cyberspace, for example via Skype.

955 Laurens van der Post says, "…the way we travel today doesn't broaden the mind. It narrows the mind…the real journey to be done is inside ourselves…" (Nancy Ryley, *The Forsaken Garden,* 31).

James G. Duncan

Global trade contributes a great deal to air, soil, and water pollution, in sending and receiving goods, and where highflying or boating tourists rule. This is a serious problem that must be remedied. We need to stay at home much more, to practice "think globally and act locally." We could have more block parties and yard sales in our towns and cities, get our hands dirty gardening, or do research on the history of our neighbourhoods.

The cultivation of our families, communities, and areas of land in our immediate vicinity, would go a long way toward correcting the widespread unhealthy nomadism that is leading us to destruction. We need to re-cultivate the relationship with sacred land that we have been ignoring, in fact, abusing, to our everlasting peril. This may be difficult at first, because we are addicted to the elixir of experiential tourism, but it will get easier as we journey on. We have only to put "localization" into practice, consciously at first, because we have been out of practice for a long while.

Alternatively, can we each develop an intimate knowledge of the *whole* Earth? No. We are not God. However, perhaps we can do so collectively. Science is a great teacher, too, and the even greater expansion of world databases will no doubt allow us to store and retrieve more information about the "way the world works," not only in an ecological sense, but possibly spiritually, too. The Earth is likely alive,[956] and therefore it may have its own spirit or soul, as Carl Jung thought.[957] Jung's reaction to an earthquake was that he "no longer stood on solid familiar earth, but on the skin of a gigantic animal that was heaving under my feet."[958]

We do not have the dominion over Nature that we seek. Such dominion, if further attempted under the ecological rules that we know

---

[956] One First Nations friend said that this is "culturally embedded." We are, says Benyus, "…a life-form on a planet that might itself be a life-form." (Janine Benyus, *Biomimicry: Innovation Inspired by Nature*, [New York: William Morrow and Company Inc., 1997], 8).

[957] Meredith Sabini, ed., 2008, 1-2, 134, 171. Jung felt that other animals have souls, too. (Sabini, 29).

[958] Ibid, 2.

On Evolution

of, would continue to be a destructive force, leading to oblivion. We are near the abyss now, and the Earth is definitely having something to say about it. She may not allow her children, sick as we are, to destroy Her, too. It remains to be seen how well we can learn the lessons of humility - and science.

Hierarchical authoritarianism has taken root over the last centuries, violently unearthing the beneficial anarchy that relied upon the self-regulating and decision-making power of the "individual in relation." This has made it acceptable to exploit land, water, sky, vegetation, and other creatures, legitimizing the rape and desecration of Motherearth. The Earth - *awakening now too* - is not taking kindly to this and is warning - e.g., through climate change and horrific natural disasters - that we may not be welcome much longer.

At the same time, the advances of science, it appears, might allow us to bring ourselves into further alignment with Nature. The new field of "biomanipulation" is promising,[959] though I would want to rename it "ecoalignment," aligning ourselves with Nature's ecological laws.[960] Thomas Berry says that we need to align ourselves with the ecological forces of the Earth,[961] and, supporting this, Ryley writes of "a healing alignment with the creative energy in the universe."[962]

"Biomimicry" is less offensive than "biomanipulation." Biomimicry, says Benyus, may be needed "not to learn *about* nature [but] to learn *from* nature...so that we might fit in...on the Earth from which we sprang."[963] Her book speaks of how we can use biomimicry to feed ourselves, to

---

[959] One example is making changes to aquatic ecosystems, mainly lakes, to offset eutrophication. (Thomas Brock, *A Eutrophic Lake: Lake Mendota, Wisconsin*, [New York: Springer-Verlag, 1985]); (James Kitchell, ed., *Food Web Management: A Case Study of Lake Mendota* [New York: Springer-Verlag, 1992]).

[960] The prefix "bio" refers to life only and does not include death, ever-present and essential to life; thus I prefer the ecological prefix "eco." Benyus describes virgin soil as smelling like "death and life commingled." (Janine Benyus, *Biomimicry*, 15).

[961] Nancy Ryley, *The Forsaken Garden*, 1998, 232.

[962] Ibid, 273.

[963] Benyus, *Biomimicry*, 9.

harness energy, to create, to improve health care, to store and use information, and to help industry and business to adapt in the New Age.

Jay Harman writes more recently, saying of biomimicry, "By constantly creating conditions conducive to life, with zero waste and a balanced use of resources, [life can be] clean, green, and sustainable."[964] Green chemistry, he says, uses a "do-no-harm principle, to develop everything from medicines to cleaning products to industrial molecules..."[965] Like Benyus, Harman cites the creation of Velcro, an early example of biomimicry. Says Harman as well, spider silk is five times tougher than Kevlar, used in bullet proof vests, and is five times stronger than steel in one measurement.[966]

Publishers Weekly commented on biomimicry on the jacket of Harman's book, saying a paint that mimics sharkskin has been used on ships to reduce friction in the water, resulting in a [possible saving] of "2000 tons of fuel per ship, per year." It also said that "...scientists are learning about anticoagulants from leeches, acoustics from dolphins, antibiotics from Komodo dragons, shock absorbers from woodpeckers, and computer networks from slime molds."

Harman sets out to further the field, writing that we are entering a new world, a "- new gold rush: biomimicry."[967] He says that "[W]hen we ignore or destroy nature," as we have been doing, "we destroy our very foundation. With our human intellect and eco-inspiration, however, we can choose to stand firmly on nature's foundation and finally flower as a species."[968] Harman boldly states "...I'm on a mission to halve the world's energy use and greenhouse gas emissions through biomimicry and the

---

964 Jay Harman, *The Shark's Paintbrush: Biomimicry and How Nature Is Inspiring Innovation*. (Ashland, OR: White Cloud Press, 2013), 8.
965 Ibid, 8.
966 Ibid, 128.
967 Ibid, 18.
968 Ibid, 19.

elimination of waste."[969] We are, he writes, near "...a golden age that is... realistically achievable."[970]

This new field has also spoken to children in a readable and timely picture book by Dora Lee, *Biomimicry: Inventions Inspired by Nature* (2011).[971] It shines a new light on the path our children could take going forward. Children will inherit many of the problems we have today, but if we can excite and interest them with biomimicry and HPMA, our future may be brighter for all.

Every field of endeavour must become "eco-conscious," to bring us back from the brink. Ecoalignment can be practiced - and is already in certain fields - ranging from ecoengineering to ecoeconomics. Useful work may be underway in emerging areas such as ecoreligion, for example, that described in Newburg's *Why God Won't Go Away*,[972] or in Abrams's *A God That Could Be Real*.[973]

Science can be, and is, used to study many mysteries. There are many secrets of the Earth that we can still, and no doubt should, uncover. More funding is needed in the "living Earth" sciences, not the space sciences. Space studies divert us from the really important things of life, our individuality in relation to family and community, ecologically-aligned with land, water, and air. But we would do well to heed H. L. Mencken's warning, noted earlier:

> Penetrating so many secrets, we cease to believe in the unknowable. But there it sits, nevertheless, calmly licking its chops.

---

969 Ibid, 3.

970 Ibid, 9.

971 Dora Lee, *Biomimicry: Inventions Inspired by Nature*, (Toronto: Kids Can Press Ltd., 2011).

972 Andrew Newburg et al, *Why God Won't Go Away: Brain Science and the Biology of Belief* (New York: Ballantine Books, 2001).

973 Nancy Abrams, *A God That Could Be Real*, 2015. Conceivably, "ecoreligion" may, for some, lead us back to a scientific paganism or nature worship.

James G. Duncan

We need to survive first, which increasingly seems an unlikely prospect; optimism is suspect. We still want to retain our unhealthy nomadism, so that we can be globetrotting businesspeople, politicians, or tourists,[974] or so that we can scientifically explore the Universe for other life that is surely there. Because of the vast distances that have to be crossed,[975] we can reach beyond our human limits with robotics. Surely the curtailment of long-distance travel is a small sacrifice to pay for our children's survival, and for the survival of other life forms on the planet.

Can there be a unity of faiths in the world, finally, of the "God-in-Nature" with the "God-outside-Nature"? I used to doubt it, but now with the progress of my own understanding of a Higher Power in First Nations and Twelve Step HPMA, I can see no reason why this unity cannot be achieved. Outside the Universe - if there is an outside - are there more Universes, or is there another "God-outside-Nature," pulling the strings of reality? I will probably never know. All that I really need to know now, with a modicum of humility, is that God loves me; of that I am certain, or I would not have survived.

We knew, before the dawn of the New Age, that we all live in one world, Mother Earth. According to Indigenous knowledge, and more and more accepted by science, She is alive, and humanity will have to respect and love Her, fully, once again. She may still have the patience to allow us to mend our ways. Other life forms are counting on us to do the right thing, both for their sake and for our own. The Earth is our *home*, not a landfill site that we can use to somehow "externalize" our garbage and toxins. *Homo sapiens* has become an amazing creature, but is concomitantly the filthiest to ever walk the planet. And the most violent: "We're the most destructive presence on Earth" says Thomas Berry, "We are the affliction of the world - its demonic presence."[976]

---

974 Long distance travel that allows for lengthy stays, years rather than days, can bring some understanding, some love, of land. Unlike Darwin, Lucas Bridges in the 1930s and Martin Gusinde in the 1940s, lived for many years in southern South America.

975 It took *five years* for a NASA probe to reach Jupiter. (CBC News, July 2016).

976 Nancy Ryley, *The Forsaken Garden*, 223.

On Evolution

The reality is that we will have to travel less, I suggest, and enjoy more fully the fruits of stable, ecological, and spiritual family and community living. Too, less nomadism will have to be accompanied by less materialism, like nomadism in that satisfaction is gained by wandering, but from store to store, not distant place to distant place.

I do not own a car and do not need one. For those who live in rural areas of the developed world, farmers, among the only ones left with a close relationship to the land, and exurbanites who commute to work in cities, cars are still a necessity. And we should honour the pedestrian more, *Homo pedester*, the "walking man," who moved, so long ago, from four legs to two, after dropping down from the trees.

Lester Brown is right. We as a species do need to unite and mobilize, as the United States did in 1941, to prepare for World War III, the defence of the planet... from ourselves.[977] The dominion of the sick "individual," of mostly European origin, and action emanating from it, has brought us as near to destruction as we dare. But as Brown says, "Saving civilization is not a spectator sport."[978] And as Marshall McLuhan felt, "There are no passengers on spaceship earth. We are all crew."

Such a worldwide effort has the potential for a re-blossoming of the Earth, perhaps a real re-creation, ultimately, of places like the Garden of Eden, or forerunners.[979] This is a useful wish, an achievable one, for a return to Rousseau's utopian "noble savage" of antiquity, "free, healthy, good, and happy[980]..., who breathes only peace and freedom; he desires

---

[977] Lester Brown, *World on the Edge* (New York: W.W. Norton & Company, 2011), 16.
[978] Ibid, 201.
[979] "Creation stories, long before the Bible was written, centred around the Goddess." (National Film Board of Canada, *Goddess Remembered*, 1989 video). Stories of Lilith elsewhere name her as Adam's equal, created, too, from dust, before the arrival of Eve - from Adam's rib.
[980] Jean-Jacques Rousseau, *A Discourse on Inequality*, London: Penguin Books, 1984, 1st published 1755, 116.

only to live and stay idle."[981] Noted earlier too, First Nations people are often described this way, exemplified by a "lazy" Jemmy Button.

Recent findings of feminist scholars show that peaceful Goddess spirituality held sway before the advent of hierarchical religions. "These Goddess-worshipping peoples were egalitarian, living for thousands of years in almost complete freedom from war and territorial conflicts as we know them today."[982]

Laurens van der Post says that "Wilderness…presents us with a blueprint…of what Creation was about in the beginning, when all the plants and animals were…fresh from the hands of whatever created them."[983] I suggest we have only to follow Higher Powered Mutual Aid, using the First Nations and Twelve Step varieties to get us started, and science to thoroughly teach us about living on our living, soulful Earth.

---

981 Ibid, 136.
982 *Goddess Remembered*, National Film Board video, 1989.
983 Nancy Ryley, *The Forsaken Garden*, 51.

Appendix D
Reflections

ACCEPTANCE

One new member of Alateen was trying to change herself, but felt the process was too slow. Someone suggested that she first think about acceptance of herself. That gave her a big lift. Changing her life was still a priority, but she now knows it will take place more naturally with further progress on just accepting herself.[984]

Of a humble acceptance, rooted in the present, Bill W. said that we must simply accept the present as it is, ourselves as we are, and others as they are,[985] before we progress to the future, one day at a time. This will allow us to develop and maintain humility before others, and before God, necessary equipment for advances in recovery.

—m—

A different kind of acceptance is about accepting the past, but not as much the present, unless one is truly humanly powerless over what is currently happening, or, God forbid, over all the awful things that *could* happen in the future. We are responsible for our own fate, for the fate of our families, communities, countries, and the planet, and must take action, today as much as we can, and certainly tomorrow, to ensure that our children and those of other species are bequeathed lives worth living.

There will be times when all we can handle is this one day: we accept it as it comes. But at others, our HPMA programs will allow us to live in the past for a while, to make amends, for example, or to live in the future,

---

[984] Al-Anon Family Groups, *ALATEEN - a day at a time*, (Virginia Beach, VA: AFG Headquarters, 1983, 168).

[985] *As Bill Sees It: The A.A. Way of Life*, 44.

thinking, planning, and acting to make our families, communities, and the planet better places for all.

―※―

> On the Road to Santiago and in life itself, wisdom has value only if it helps to overcome some obstacle.[986]

The Serenity Prayer asks for the serenity to accept things that we cannot change, courage to change what we can, and the wisdom to differentiate between them. Many obstacles seem humanly immovable, such as hubris. We are therefore led, if we have gained sufficient humility, to the wise acceptance of some obstacles, not their overcoming. Large egos may sometimes be accepted, even as we diligently practice deflating them to right-sized, manageable proportions, before God and others, through HPMA programs.

Alcohol

A substance that is "cunning, baffling, powerful!"[987]

It has been said that if alcohol was discovered today, it would be a banned substance. Thus, we might elevate, from suggestion to recommendation, that alcohol should be made less available than it is today.

Alcohol's use "is as old as civilization itself," says Austin.[988] Standage tells us that beer, wine, and spirits are the first three of six beverages that "chart the flow of human history," the rest, coming later, being coffee, tea, and cola.[989] Unfortunately, beer, wine, and spirits are still far too popular, around the world.

---

[986] Paulo Coelho, *The Pilgrimage: A Contemporary Quest for Ancient Wisdom*, translated by Alan Clarke (New York: HarperCollins Publishers, 1995), 162.

[987] *Alcoholics Anonymous* (The Big Book), 58.

[988] Gregory Austin, *Alcohol in Western Society from Antiquity to 1800: A Chronological History* (Santa Barbara, CA: ABC-CLIO Information Services, 1985), xiii.

[989] Tom Standage, *A History of the World in 6 Glasses*, 2.

On Evolution

Uncorking the Past: The Quest for Wine, Beer, and Other Alcoholic Beverages, a book by Patrick McGovern, has its first chapter titled, *Homo Imbibems: I Drink, Therefore I Am*. He writes that radio wave scans have shown that alcohol is present in interstellar space and surrounding new star systems. Citing the view of one biologist, he says, "...alcoholism among humans is rooted in the evolutionary history of primates," the "drunken monkey hypothesis."[990] McGovern traces alcohol consumption to the Neolithic, starting around 8000 BC, and to the Paleolithic, starting many thousands of years earlier.[991]

An April 4, 2011, study released in the *Journal of the Canadian Medical Association* said that, despite its profound serious effects on health, "Alcohol is often not 'on the radar screen' as a major public health issue." We are quite unaware, perhaps wilfully. Alcohol and civilization have indeed walked hand in hand down the aisle that is leading to a failed marriage.

Fortunately, we already have in place a simple, planet-wide veneer of spirituality that can be used to return humanity to sobriety, sanity, and serenity - the Higher Powered Mutual Aid Twelve Step movement that complements ancient First Nations HPMA. As a suggestion, First Nations and Twelve Step spirituality - where all can have the God of our own understanding[992] - is what we must seek in the new millennium and beyond.

However, alcoholism and other addictions are still common, despite decades-long battles by A.A., Al-Anon, and other Twelve Step organizations. So, might we develop a contingency plan for "attraction

---

990 Patrick McGovern, *Uncorking the Past: The Quest for Wine, Beer, and Other Alcoholic Beverages*. (Berkeley, CA: University of California Press, 2009), 1-8.

991 Ibid

992 Kasl doubts, in error, that a Higher Power can be anything one wants it to be. (Charlotte Kasl, *Many Roads, One Journey*, 1992, 146).

rather than promotion" based on the appeal to practice Twelve Step principles in all our affairs? Three strategies might be used here, namely:

- Bringing our lives out of the closet to display personal success in dealing with our addiction problems and to see it as the fortune that we can fruitfully share with others.[993]
- Forming outside groups or organizations that address family, community, national, and global problems.
- Beginning a campaign to change Tradition Eleven, to allow Twelve Step societies and members to use promotion rather than only attraction. Because the Twelve Step movement has stalled, I feel that the pioneers would approve.

Anonymity

According to one alcoholic who knew him well, as already mentioned, Dr. Bob felt that we could break the anonymity tradition in two ways: by offering one's name to the press or radio; or by being so nameless that one can't be found by other alcoholics.[994]

One A.A. member commented on the issue in A.A.'s early years, saying that he had a one-hundred-name address book, complete with first and last names, telephone numbers, and addresses. He felt that fellow members in A.A. had later gotten "carried away" with anonymity, prompting him to say that anonymity "gets to be a joke."[995]

To Joe P., Dr. Bob was convinced that alcoholics should let their communities know of their membership in A.A.[996] This would be "carrying the message" by firm example, with the possibility of attracting other alcoholics to A.A., and a step forward in removing the stigma associated with alcoholism.

---

993 *Alcoholics Anonymous* (The Big Book), 123–24.
994 *Dr. Bob and the Good Oldtimers*, 264.
995 Ibid, 265.
996 Ibid

On Evolution

Dr. Bob and Bill W. were on the same anonymity page. Bill W. said that when being helpful presented itself, the alcoholic found he could talk about his problems with "almost anyone."[997] He said as well, "...no A.A. [member] need be anonymous to family, friends, or neighbors. Disclosure there is usually right and good. Nor is there any special danger when we speak at group or semipublic A.A. meetings, provided press reports reveal first names only."[998] This would also be of great assistance in removing the stigma of alcoholism and in spreading the good news. *As Bill Sees It* says that only at the "top public level" is anonymity needed.[999]

An article in the *New York Times* said that, "More and more, anonymity is seeming like an anachronistic vestige of the Great Depression, when A.A. got its start and when alcoholism was seen as not just a weakness but a disgrace." It quotes one professor as saying, "it's extremely healthy that anonymity is fading." A novelist said in the article, too, "It seems crazy that we can't just be out with it, in this day and age," adding, "I don't want to hide my sobriety; it's the best thing about me."[1000]

The article further quotes author Susan Cheever as saying:

We are in the midst of a public health crisis [and] A.A.'s principle of anonymity may only be contributing to general confusion and prejudice.

Cheever added:

I am increasingly uncomfortable with this level of dishonesty. This dancing around and hedging, figuring out ways of saying it that aren't really saying it, so that people in recovery know what

---

997 *As Bill Sees It*, 43.
998 *Alcoholics Anonymous Comes of Age*, 293
999 *As Bill Sees It*, 43.
1000 *New York Times*, May 6, 2011.

I am talking about - all the code words. I am sure this is not what Bill [W.] intended.

—⚋—

The Al-Anon Family Groups, and its offshoot, Alateen, often cannot be reached by people needing help because of someone else's drinking. Many leading dysfunctional lives do not know that they have been affected. Most know of the famous A.A. but not the Al-Anon Family Groups, even in North America, over sixty years after its founding in 1951. Bill W. predicted that Al-Anon would have five times as many members as A.A.[1001] but this has not yet happened. More attraction is needed by carrying the message in all our affairs, and by the so-far prohibited promotion of Twelve Step programs.

—⚋—

Authority

In his mention of the "gentle Russian prince" and the "benign anarchy" of A.A., Bill W. spoke of the wonderful freedom from authoritarianism that can be found in A.A., where no one person is in charge.[1002]

Coelho often refers to a human master.[1003] But, the concept of a human master is ecologically and spiritually unsound, since authority over the self should come from the living light/energy, or soul/spirit, innate within us, not from other human sources. This talent is *self-responsibility*. Healthy decision-making for "individuals in relation" always starts from within.

1001 *First Steps: Al-Anon...35 years of beginnings*, (Virginia Beach, VA: AFG Headquarters, Inc.) 61.
1002 *Alcoholics Anonymous Comes of Age*, 224-25.
1003 Paulo Coelho, *The Pilgrimage: A Contemporary Quest for Ancient Wisdom*, 1995.

On Evolution

Does God love me?
Yes. "You are a child of the universe, no less than the trees and the stars; you have a right to be here . . ."[1004]

Garden of Eden
Often quite skeptical of the existence of an actual Garden of Eden, I did not like the gender, other animal (i.e., the talking snake), and unscientific biases in the Bible's book of Genesis. However, something did appeal about the Garden: the bounty of nature and the laid-back lifestyle of Adam and Eve. They did not appear to have much to do to survive and, perhaps as a result, like unsupervised children, they got into trouble.

Earlier I said that we might make a worldwide effort for "a re-blossoming of the Earth, perhaps a real re-creation . . . of places like the Garden of Eden." As noted earlier, Rousseau tended in this direction in his *A Discourse on Inequality*. First Nations Grandfather, Rolling Thunder, says this[1005] and so does Ronald Wright in *A Short History of Progress*, both as also noted.[1006]

I also previously noted that Jemmy Button, a Fuegian captive/hostage/guest of the British in the Falkland Islands,[1007] might have been seen as "lazy" because work was foreign to him. The European work ethic had not been instilled. Bridges notes of the Ona, humorously, that the men "had never done a steady day's work in their lives, and if left alone accomplished next to nothing..."[1008]

1004 From *Desiderata*, a beautiful and inspiring prayer by Max Ehrman in about 1920.

1005 Karen Speerstra, *The Green Devotional: Active Prayers for a Healthy Planet*. (San Francisco: Conari Press, 2010), 244.

1006 Ronald Wright, *A Short History of Progress*, 9.

1007 De Paolo is uncertain whether the Fuegians were captives, hostages, guides, interpreters, or guests, and thinks that combinations of these were possible. (*The Ethnography of Charles Darwin*, 54–56.)

1008 Lucas Bridges, *Uttermost Part of the Earth*, 216.

James G. Duncan

The lands of Indigenous peoples were bounteous before the coming of the white man,[1009] so much so that they would not have had to work. John Mohawk writes that when whites arrived in the Americas, suddenly they had plenty to eat, escaping the hunger of the Old World.[1010] Sioui refers to a "startling and meticulous description of the food resources available to Natives [of Florida] [in] "Native American Paradise Lost.""[1011] As earlier noted, said Sauk and Fox leader, Black Hawk:

> We always had plenty; our children never cried from hunger, neither were our people in want…Our village was healthy…If a prophet had come to our village in those days and told us that the things were to take place which have since come to pass, none of our people would have believed him.[1012]

Was much of the Earth, untouched by European hands, a kind of Garden of Eden, where people could simply gather and hunt, kick back, enjoy life more easily, and leave a small ecological footprint? Likely. Ryley writes of the re-construction of "that larger garden, the planet", [1013] and, in her concluding chapter, of the "Return to the Garden."[1014]

The Garden of Eden may have existed, if not in all the details, then in the general idea of a land of plenty.

1009 Miriam Simos, *The Earth Path*, (San Francisco: Harper Collins Publishers, 2004), 10.
1010 John Mohawk, "Subsistence and Materialism," (Jerry Mander et al, eds., *Paradigm Wars*, 2006, 27).
1011 Georges Sioui, *For an Amerindian Autohistory*, 9.
1012 Karen Speerstra, *The Green Devotional*, San Francisco: Conari Press, 2010, 132.
1013 Nancy Ryley, *The Forsaken Garden*, 1998, 275.
1014 Ibid, 273-79.

On Evolution

To grow in the midst of dangers is the fate of the human race, because it is the law of the spirit."[1015]

Dubos's growth "in the midst of dangers" conforms to the Darwinian "law" of the unrelenting, competitive "struggle for existence." However, the true "law of the spirit" is found in serenity-producing, ecological Higher Powered Mutual Aid, the way of life that we need to adopt again, to get us back to the Garden, after 3,000 years away from it.

Out of the pursuit of the individual good and the common welfare in HPMA practice, will come the planetary good, by definition getting us back on track as a species in our daily lives, leading us to the Garden once again.

In a cave in southwestern France, one of the oldest existing sculptures of the human form was found. Archaeologists date her creation to 35,000 B.C.[1016]

Accomplished Canadian astronaut Chris Hadfield dates another carving found in a cave, a flute made of bone, at 45,000 B.C.[1017]

Archaeologists and anthropologists are beginning to discover evidence of a widespread early civilization based on cooperation. These goddess-worshipping peoples were egalitarian, living for

---

1015 Rene Dubos, *Mirage of Health*, 1959, 282.
1016 *Goddess Remembered*, video, 1989, National Film Board of Canada.
1017 Chris Hadfield, "The Ottawa Folkfest," Hogsback Park, September 2013.

thousands of years in almost complete freedom from war and territorial conflict as we know it today.[1018]

Happiness

[T]he search for happiness is a personal search and not a model we can pass on to others.[1019]

I disagree. The search for happiness is not only a personal search, but a societal one too, with an excellent model to live by, to help each other in the search: Higher Powered Mutual Aid rooted in the immutable laws of Nature.

How is happiness created? In my estimation through:

- exercising the responsibility, right, and duty, to fulfill one's God-given purpose and potential, each within the healthy ego's capabilities.
- the ability to "right-size" ourselves as human, not super- or sub-human.
- having a real love of the healthy ego, self-love.
- having loving family and community structure. Without rooted, stable families, groups, and communities, insecurity, fear, and mistrust arise, and abound, for all.
- understanding and employment of the "individual in relation" to the common welfare, and its social and ecological implications.
- opposition to the competitive, destructive, fictitious "individual," rather than the cooperative, creative, common-welfare-conscious "individual in relation."
- de-emphasising competition that fosters the growth of the "I'm-better-than-thou" ego, rather than learning the HPMA principles of humility and cooperation.

1018 *Goddess Remembered*, video, 1989, National Film Board of Canada, back cover.
1019 Paulo Coelho, 1995, 257.

On Evolution

Humour

A.A. members often have wonderful humour, likely taking after Bill W. and Dr. Bob. I read a story about one man's struggles with alcohol; it goes something like this:

> A man started a drinking ritual with his two brothers, who now lived on different continents. They agreed to drink one pint for each one, for old time's sake. When one evening he ordered 2 pints instead of the usual 3, the bartender and patrons thought that one of his brothers had died. When he offered his condolences, the man said: "Oh, no, everyone is fine. I just joined the Mormon Church, and I had to quit drinking."[1020]

Former leader of the U.S.S.R., Nikita Khrushchev, chided wealthy U.S. industrialist Armand Hammer, that he could not take his fortune with him when he died. Hammer is said to have replied, "Well, if I can't take it with me, I'm not going."

Illness

> [T]he dominant global economic system can thrive only by constant economic expansion, constantly expanding exploitation of scarce resources, constantly expanding consumerism, privatization of all elements of the natural commons, export-oriented production - involving ecologically disastrous long-distance shipping - and the homogenization of global cultures within a commodified, commercialized, and yet individualistic…worldview. It is a system that measures

---

1020 Thomas Cathcart et al, *Plato and a Platypus Walk into a Bar…* (New York: Penguin Books, 2007), 29.

its success purely by the achievement of short-run economic goals. That is the formula that is killing the world.[1021]

—∞—

The choice is ours - yours and mine. We can stay with business as usual and preside over an economy that continues to destroy its natural support systems until it destroys itself, or we can be the generation that changes direction, moving the world onto a path of sustained progress. The change will be made by our generation, but it will affect life on Earth for all generations to come.[1022]

—∞—

James Lovelock put forward the idea that the Earth is a self-regulating, possibly living entity.[1023] For many Indigenous peoples, this knowledge is "culturally embedded."[1024] If this is so, the Earth may now be finally fed up with the human species - with its anti-ecological war against it - and is itself trying to survive. It may be signalling that we must mend our ways so we may rejoin Earth's community.

Eckhart Tolle gets to the heart of the matter with his discussion of planetary "madness",[1025] our fondness for war at the macro and micro levels, against ourselves, against "other life forms, and the planet itself . . ."[1026]

1021 Jerry Mander et al, eds., *Paradigm Wars*, 2006, 226.

1022 Lester Brown, *World on the Edge*, (New York: W.W. Norton & Company, 2011), 201-202.

1023 James Lovelock, *The Ages of Gaia*, 1990. We saw earlier that Carl Jung had similar sentiments. (Meredith Sabini, ed., 2008, 1-2, 134, 171).

1024 Personal communication.

1025 Eckhart Tolle, *A New Earth: Awakening to Your Life's Purpose*. (New York: Dutton, 2005), 8-12.

1026 Ibid, 11.

On Evolution

Human history, he says accurately, ". . . is to a large extent a history of madness."[1027]

INDIGENOUS PEOPLES (INCLUDING FIRST NATIONS)

Indigenous societies...have traditionally shared an opposite set of paradigms and values from those of the larger society...these include: shared natural commons; collective land ownership; philosophies of reciprocity, exchange, sharing and balance; religions integrated with nature; the primary virtue of sacred ancestral lands; consensus decision-making on economic and political matters; and the long-term viability of locally based, self-sufficient, subsistence-oriented economies. Most of all there is a nearly universal commitment to social, political, economic, and spiritual values that sustain strong, coherent land-based communities.[1028]

[W]e need to celebrate the clear reality that Indigenous societies are a living alternative to the current economic, political, and philosophical models of our time - and a successful one.[1029]

[I]t is obvious to me that whatever [we] can do to further the Indigenous cause, in its many struggles and forms, also furthers our own and our children's.[1030]

In this new age of electronic communications, non-Indigenous activists are . . . finding common cause with Indigenous struggles, especially on issues of environmental sustainability, human rights, democracy, equity, self-determination, and sovereignty . . . Collective action is [still] possible.[1031]

1027 Ibid
1028 Jerry Mander et al, *Paradigm Wars*, 2006, 226-27.
1029 Ibid, 226
1030 Ibid, 227
1031 Ibid, 224.

Model/Structure

The Twelve Step pioneers uncovered "a way to God," the Higher Powered Mutual Aid model for benign anarchy, to aid in healing, a method long known in First Nations. This is power from within us and from the God of our own understanding.

One A.A. leaflet succinctly states that it helps alcoholics without human authority.[1032]

The *A.A. Service Manual* describes the "bottom-up" structure of A.A. in a chart, with the groups paramount in the decision-making process.[1033] As was noted earlier, Nowinski emphasizes the "bottom-up" structure of A.A. organization.[1034]

The *Al-Anon Alateen Service Manual* contains a similar "bottom-up" chart depicting Al-Anon groups as paramount in the decision-making process.[1035]

Rene Dubos had this to say on the subject of model/structure:

> [U]nless men become robots, no formula can ever give them [health and happiness] permanently . . . nor can their societies achieve a structure that will last for millennia . . .[1036]

1032 Alcoholics Anonymous, "Circles of Love and Service: How the Parts of A.A. All Fit and Work Together," pamphlet, (New York: A.A. World Services, Inc., 1978).

1033 Alcoholics Anonymous, *The A.A. Service Manual, Combined with the Twelve Concepts for World Service by Bill W.*, 2010–2011 edition. Reflecting Actions of the 2010 General Service Conference (New York: A.A. World Services, Inc., 2010), S16.

1034 Joseph Nowinski, *If You Work It, It Works!* 2015, 4, 13.

1035 Al-Anon Family Groups, *Al-Anon Alateen Service Manual*, 2017-2019 edition (Virginia Beach, VA: AFG Headquarters, Inc., 2010), 158.

1036 Rene Dubos, 1959, 278.

On Evolution

The exception is the survival of Indigenous peoples, such as ancient First Nations, with their knowledge and practices handed down despite centuries of banned ceremonies and languages. Dubos seems to have known nothing of the great potential of the creative, healing structures of HPMA of First Nations and Twelve Step societies, a way of living that can continue, again, for as long as the sun shines.

PRACTICE
We have been out of HPMA practice in spiritual and ecological thinking, acting, speaking, listening, being, belonging, and becoming, for millennia. No wonder we are in such trouble. Change may be slow, but we must boldly, and soon, advance it with HPMA.

In re-creating our "individual in relation" selves, self-regulating self-responsibility - an innate law - comes first, not power imposed from without, authoritarianism. Armed with this understanding, we can practice carrying out effective, responsible action in our groups and in the world around us.

The spiritually and ecologically "fit" are those who cooperate best and who have arisen through, and who continue to practice, First Nations and Twelve Step forms of Higher Powered Mutual Aid.

> You learn to speak by speaking, to study by studying, to run by running, to work by working, and just so you learn to love God and man by loving.[1037]

1037 Francis de Sales quoted by Bishop Jean-Pierre Camus in *The Spirit of Saint Francis de Sales*, first published 1641 (Boston, MA: IndyPublish), 2004.

Just so, listening to God takes practice. True listening should start early in life, but rampant human authoritarianism often prevents it.

POTENTIAL AND PURPOSE

Twelve Step sponsors must be able to take the hand of a newcomer to help her/him reach inside for the immanent courage and faith that exists in all of us. Let us help the newcomer to produce another miracle.[1038]

One Al-Anon member wanted to know why she acted the way she did, and said what she said. Provided with this knowledge through participation at HPMA meetings, she helped herself to become the type of person she really was. It will also give her good ideas about doing a Fourth Step inventory, to improve herself and her life.[1039]

Working the Fifth Step, admitting faults to ourselves, to God, and to another person, has helped us to become the wonderful human beings that we really are.[1040]

The Twelve Steps assisted us to become the persons we were meant to be. As we began to know ourselves, with courage and sincerity, admitting our flaws and our talents, we built on the good aspects of our lives and sliced away the bad. Now we are free to be rid of the self-deceptions that held us back from truly knowing ourselves.[1041]

—⁂—

We must let go of "the need to know who [we] are."[1042]

[1038] *One Day at a Time in Al-Anon*, (Virginia Beach, VA: AFG Headquarters, Inc., 1968), 94.
[1039] Ibid, 101.
[1040] Ibid
[1041] Ibid, 117.
[1042] Eckhart Tolle, 90.

On Evolution

No. The search for "who we are" is in fact necessary and integral to HPMA purpose - beginning in the Fourth Step, making "a searching and fearless moral inventory of ourselves." This is a prerequisite to becoming the best that we can be, with the responsibility, the right, and the duty, to reach toward our God-given potential and purpose.

I want to be all that I am capable of becoming, so that I may be . . . there is only one phrase that will do - a child of the sun.[1043]

We are all children of God, each in some way, with great potential and purpose. But, as all HPMA practitioners know, there is no escaping the natural truth, and that is "all that I am capable of becoming" is carried out within the common welfare.

[Y]ou cannot become successful. You can only be successful.[1044]

No. Success in the Twelve Step HPMA process begins with the onset of self-nurturing potential. If you are fortunate, you may be brought up in a culture that permits the pursuit of potential from childhood.

In the course of healing through the practice of First Nations and/or Twelve Step HPMA, we can learn our purpose and can then act upon it to achieve potential.

Hornbacher says that A.A. promotes the process of achieving one's potential in life, one of the chief benefits of being a Twelve Stepper.[1045] It must be said that all, believers and non-believers, can fit themselves into this healthy, potential-serving program.

1043 Rene Dubos, 1959, 281
1044 Eckhart Tolle, 270
1045 Marya Hornbacher, *Sane: Mental Health, Addiction, and the Twelve Steps*, 82.

James G. Duncan

Qualification

"AA for the Native North American: Trails to Freedom," pamphlet, New York: Alcoholics Anonymous World Services, Inc., 1989.

"A.A. and the Gay/Lesbian Alcoholic," pamphlet, New York: Alcoholics Anonymous World Services, Inc., 1989.

"Al-Anon is for…African Americans…African Canadians…All People of Color", pamphlet, Virginia Beach, VA: Al-Anon Family Group Headquarters, Inc.,

"Al-Anon is for Gays and Lesbians," pamphlet, Virginia Beach, VA: Al-Anon Family Group Headquarters, Inc.

If you say you're a member, you're a member.

Right and Wrong

I am right; you are wrong is one of the ways in which the ego strengthens itself.[1046]

Tolle is led, from his question "Is there no such thing as right or wrong?"[1047] to the faulty conclusion, "Nobody is wrong."[1048]

There *is* in fact right in the world, and wrong, too. We know that it is good "to love others as yourself." At least we should know it, if we are practicing our HPMA programs.

Tolle falls into the trap, like so many others, of misinterpreting Shakespeare's *ironic* phrase in the play, *Hamlet*:

There is nothing either good or bad but thinking makes it so.[1049]

There *is* good and bad in the world, a conclusion from clear thinking. Simply put, hating is bad, of course, while loving your neighbour as yourself is good.

Quoting from Zen Buddhism, Tolle says:

1046 Eckhart Tolle, 69

1047 Ibid

1048 Ibid, 76

1049 William Shakespeare, *Hamlet*, Act 2, Scene 2.

On Evolution

Don't seek the truth. Just cease to cherish opinions.[1050]

Zen, then, too, is wrong on this score. We all have access to humanly knowable truth - and therefore solid opinions and convictions - rooted in our moral, natural Earth.

Tolerance of all behaviours and habitual permissiveness are diseases that result from the attempted destruction of the "individual in relation." Global culture, instead, puts forward the simplistic notion of the fictitious "individual" who, sick as (s)he is, is under the illusion that (s)he is somehow separate from other creatures and Nature.

In Twelve Step programs, we use the slogan "Live and Let Live" to erode our unhealthy egos and to accept others as they are. However, as with most things, this can be carried too far, perhaps to unhealthy extremes. Sexual assault, ethnic cleansing, racism, and divisiveness, for example, must end.

Bill W. said that there are things - some convictions for example - on which we cannot compromise. He calls for careful discrimination.[1051]

RIGHT-SIZING EGO

Bill W. felt that "right-sizing ego"[1052] was key, not only for alcoholics, but for all of humanity. Science may not be able to solve all mysteries

---

1050 Tolle, 2005, 121

1051 *As Bill Sees It*, 59

1052 Bill W. also described this as "deflation at depth." (*Alcoholics Anonymous Comes of Age*, 64), and "ego deflation at great depth" (*Bill W.: My First 40 years*, 154).

and human intellect should not be our new God. Humility, he felt, was a greater virtue than intellect.[1053]

In *Sane: Mental Health, Addiction, and the Twelve Steps*, Marya Hornbacher describes the difference between healthy self-esteem, and arrogance or egotism.[1054] "Right sizing" our egos, she says, should be addressed through humility and maturity in relation to others and to our Higher Powers.[1055]

Charles Darwin wrote that humans have "a god-like intellect."[1056]
Did he consider his powerful intellect "god-like"? For Darwin, "right-sizing" ego-deflation may have been needed.

> There is no reason to doubt . . . the ability of the scientific method to solve each of the specific problems of disease . . . all the separate problems of human health can and will eventually find their solution.[1057]

Is this our mortal enemy, hubris? There is reason to doubt this now; Dubos says:

> Whereas other living things survive through adaptive changes in their bodies and their instincts, man strives to impose his

---

1053 *Twelve Steps and Twelve Traditions*, 30; *As Bill Sees It*, 60.
1054 Marya Hornbacher, 27.
1055 Ibid, 41-42.
1056 Charles Darwin, *The Descent of Man*, 613.
1057 Rene Dubos, 1959, 26.

On Evolution

own directional will on the relations he has with the rest of the world.[1058]

Following our own "directional will" often *does* get us into trouble. Dubos seems to be speaking of the mostly European-derived, inflated-ego societies, not the Higher Powered Mutual Aid of First Nations or Twelve Step societies that produce the "right-sized ego."

Modern man believes that he has achieved almost complete mastery over the natural forces which molded his evolution in the past and that he can now control his own biological and cultural destiny. But this may be an illusion. Like all other living things, he is part of an immensely complex ecological system and is bound to its components by innumerable links . . . At some unpredictable time and in some unforeseeable manner, nature will strike back.[1059]

From [scientists] [humanity] has acquired the faith - or rather the illusion - that society can be planned in a manner that will assure plenty, health, and happiness for everyone and thus solve all the great problems of existence.[1060]

When Dubos wrote this book, these two statements were far seeing.

—⚏—

[H]uman behavior is governed not only by biological necessities but also the desire for change [for more].[1061]

1058 Ibid, 266
1059 Ibid, 266–67
1060 Ibid, 270
1061 Ibid, 267

Rather, it is the large-ego, illusionary, insecure, and fearful "individual" that feeds the excessive desire for change, and ironically, for more connection. Disconnected from the Earth, we must seek our "individual in relation" to the common welfare, to become happy, healthy, and secure.

—⚭—

The problem that "every human being carries within [is] the ego."[1062]

In the shift that we need to undertake, "the ego is destined to dissolve."[1063]

Here, Tolle may be speaking of ridding ourselves of the *unhealthy* ego, self-aggrandizement, the ego that makes us think that we are superior beings destined for the stars, not tied to the Earth as humble creatures of God. Tolle's concern then seems to be the *overly large* ego that is spoken of in Twelve Step societies, where "ego-deflation," not "ego-abandonment," is a key part of recovery.

However, Tolle's *dissolution* of the ego amounts, ultimately, to abandonment of the self. Tolle may not mean this, but his idea needs clarification. It is only in a few places that he mentions the *unhealthy* ego, the ". . . yourself as better than . . . others," feeling ". . . bigger, superior," or the ". . . feeling of superiority."[1064] We must be careful not to throw the baby - the healthy, "right-sized ego" - out with the "I-am-better-than-thou" bathwater, lest we abandon ourselves and have nothing left to work with.

The wholesale abandonment of the ego would bring about the collapse of humanity, just as surely as will the triumph of the unhealthy ego. Moderation is key.

Recovery of the "right-sized" ego, a re-powerment, is more difficult for adults. It is much easier if we are allowed to develop the "right-sized" ego from childhood, as it has been in ancient First Nations societies.

1062 Eckhart Tolle, 2005, 13

1063 Ibid, 19

1064 Ibid, 61-66.

On Evolution

In the eyes of the ego, self-esteem and humility are contradictory.[1065]

The "right-sized" human ego can have both healthy self-esteem *and* humility, as we know from experience in practicing HPMA. If the ego is "right-sized," "individual-in relation-sized," HPMA practitioners can be at peace, content, in realistic relations with others who also have healthy, right-sized egos. Some egos are too large, others too small, but they can all be accepted first, then "right-sized" by practicing HPMA.

—⚜—

"Right-sizing" our egos, in the practice of Higher Powered Mutual Aid, may help us to modify - or even end - the insatiable desire for more.

> Once his essential biological needs are satisfied, man develops other urges which have little bearing on his survival as a species..., man desires change for change's sake, without regard to any biological need.[1066]

It is the dominant, "addicted to more" global civilization that always has other urges, not in, for example, Twelve Step or First Nations HPMA cultures.

> It is important . . . that there be available opportunities for change, for when they are lacking, man is apt to satisfy his thirst for change by acts of violence or destruction.[1067]

Dubos is generalizing here from the European-derived "addicted to more" world culture, not peaceful, serene First Nations and Twelve Step people in HPMA programs.

—⚜—

1065 Ibid, 109.
1066 Rene Dubos, 275.
1067 Ibid, 276.

Who are the meek or the humble, and what does it mean that they shall inherit the Earth? . . . The meek are the egoless.[1068]

The meek and humble, rather, are those with human-sized, "right-sized" egos, not those entirely without ego, the living dead.

Wisdom

"The true path to wisdom can be identified by three things."

First, "agape", spiritual love, especially as distinct from erotic or brotherly love.

Second, "[I]t has to have practical application; otherwise wisdom becomes useless, like a sword that is never used."

Third, "[I]t has to be a path that can be followed by anyone . . ."[1069]

On all counts, HPMA programs qualify as paths to wisdom.

—⁂—

> It is often suggested that a moratorium on science would give mankind the opportunity to search its soul and discover a solution to the problems that threaten its very survival.[1070]

With an adequate opportunity to extend HPMA programs to all peoples, worldwide, humanity can come up with "a solution to the problems that threaten its very survival," an answer to our many crises.

Dubos was optimistic - in 1959 - saying:

> [D]espite so many disheartening setbacks, the activities of man seem to have on the whole a direction upward and forward which tends to better his life physically, intellectually, and morally.[1071]

1068 Eckhart Tolle, 309
1069 Paulo Coelho, 27.
1070 Rene Dubos, 272.
1071 Ibid, 277. I don't think he would say the same thing in 2017.

On Evolution

He qualified this on the same page, adding:

> [C]ertain of man's ideals and goals threaten to have consequences unfavourable for the human species..., the very mastery of nature may release dangers that cannot be controlled.
>
> Men will develop new urges, and these will give rise to new problems, which will require ever new solutions. Human life implies adventure, and there can be no adventure without struggles and dangers."[1072]

Sure, but we do not have to go to such extremes, by always wanting more of everything, never satisfied with enough. Our urges need to be tempered, some of them ended - the dangerous ones - by participating in HPMA programs that can teach us humility, morality, sobriety, sanity, serenity - and frugality.

Dubos says, probably close to the truth:

Awareness of dangers is not likely to deflect the course of mankind . . . [1073]

Indeed. Our involvement in the pursuit of "more of everything" has led us to profound dangers. Unawares, or perhaps uncaring, we continue living as we do. Bill W. said: "...spiritual rebirth may be the only alternative to extinction."[1074]

—⚏—

The story of an unfortunate frog comes to mind. Suddenly immersed in hot water by some tyrant, the frog will jump out. However, if s(he) immerses the frog in water that is only gradually made hotter, the frog will not jump out.

The water is getting hotter. Will we jump out or stay immersed?

1072 Rene Dubos, 278

1073 Ibid, 277

1074 *Alcoholics Anonymous Comes of Age*, 231.

Appendix E
Permissions

Diligent efforts were made to obtain permissions from copyright holders. The author and publisher are grateful for the use of all excerpted material.

The Twelve Steps, Twelve Traditions, and brief excerpts from the texts *'PASS IT ON'* and *Alcoholics Anonymous Comes of Age* are reprinted with permission of Alcoholics Anonymous World Services, Inc. ("AAWS"). Permission to reprint the Twelve Steps, Twelve Traditions, and brief excerpts does not mean that AAWS has reviewed or approved the contents of this publication, or that A.A. necessarily agrees with the views expressed herein. A.A. is a program of recovery for alcoholism *only* - use of the Twelve Steps and Twelve Traditions in connection with programs and activities that are patterned after A.A., but address other problems, or in any other non-A.A. context, does not imply otherwise.

Excerpts for the following are copyrighted by Al-Anon Family Groups Headquarters, Inc. Reprinted by permission of Al-Anon Family Group Headquarters, Inc.

Al-Anon Declaration; Al-Anon's Third Tradition; Al-Anon's Twelfth Step; *Living with an Alcoholic,* copyright 1966; *Lois Remembers*, copyright 1979; *Al-Anon's Twelve Steps and Twelve Traditions*, copyright 1981; *Lois and the Pioneers*, copyright 1982; *As We Understood...*, copyright 1985; *First Steps - Al-Anon . . . 35 Years of Beginnings*, copyright 1986; *Paths to Recovery - Al-Anon's Steps, Traditions, and Concepts*, copyright 1997; *Having Had a Spiritual Awakening...*, copyright 1998; and *The Al-Anon Family Groups - Classic Edition*, copyright 2000.

The excerpts from Al-Anon Conference Approved Literature are reprinted with permission of Al-Anon Family Groups Headquarters, Inc. Permission to reprint these excerpts does not mean that Al-Anon Family Groups Headquarters, Inc. has reviewed or approved the contents

James G. Duncan

of this publication, or that Al-Anon Family Groups Headquarters, Inc., necessarily agrees with the views expressed herein. Al-Anon is a program of recovery for families and friends of alcoholics; use of these excerpts in any non-Al-Anon context does not imply endorsement or affiliation by Al-Anon.

Bibliography

ALCOHOLICS ANONYMOUS

The A.A. Grapevine. *AA around the World: Adventures in Recovery.* New York: The AA Grapevine, Inc., 2000

The A.A. Grapevine. *Best of the Grapevine.* New York: The AA Grapevine, Inc., 1985.

The A.A. Grapevine. *The Language of the Heart.* New York: The AA Grapevine, Inc., 1988.

A.A. in Prison. New York: Alcoholics Anonymous World Services, Inc., 2003.

The A.A. Service Manual. Combined with Bill W. (1962) *Twelve Concepts for World Service*, 2010–2011 Edition. New York: Alcoholics Anonymous World Services, Inc., 2010.

Alcoholics Anonymous (The Big Book). Originally published 1939, 3rd edition. New York: Alcoholics Anonymous World Services, Inc., 1976.

Alcoholics Anonymous (The Big Book). Originally published 1939, 4th edition. New York: Alcoholics Anonymous World Services, Inc., 2001.

Alcoholics Anonymous Comes of Age: A Brief History of A.A. New York: Alcoholics Anonymous World Services, Inc., 1957.

As Bill Sees It: The A.A. Way of Life. Selected writings of A.A.'s cofounder. New York: Alcoholics Anonymous World Services, Inc., 1967.

Came to Believe . . . New York: Alcoholics Anonymous World Services, Inc., 1973.

Daily Reflections: A Book of Reflections by A.A. Members for A.A. Members. First published 1990. New York: Alcoholics Anonymous World Services, Inc., 2009.

Dr. Bob and the Good Oldtimers: A Biography, with Recollections of Early A.A. in the Midwest. New York: Alcoholics Anonymous World Services, Inc., 1980.

Experience, Strength and Hope. New York: Alcoholics Anonymous World Services, Inc., 2003.

'PASS IT ON': The story of Bill Wilson and how the A.A. message reached the world. New York: Alcoholics Anonymous World Services, Inc., 1984.

Twelve Concepts for World Service. Bill W. As adopted by the Twelfth Annual General Service Conference of Alcoholics Anonymous. New York: Alcoholics Anonymous World Services, Inc., 1962.

Twelve Steps and Twelve Traditions. New York: Alcoholics Anonymous World Services, Inc., 1953.

AL-ANON FAMILY GROUPS

Al-Anon Alateen Service Manual, 2015-2018. Virginia Beach, VA: Al-Anon Family Groups Headquarters, Inc., 2015.

Al-Anon Faces Alcoholism. First printing. Virginia Beach: AFG Headquarters, Inc., 1965.

Al-Anon Faces Alcoholism, 2nd edition. Virginia Beach: AFG Headquarters, Inc., 1992.

The Al-Anon Family Groups: Classic Edition. Original edition with footnotes and annotations added in 2000. Virginia Beach: AFG Headquarters, Inc., 2000.

On Evolution

Al-Anon's Twelve Steps and Twelve Traditions. 9th printing. Virginia Beach: AFG Headquarters, Inc., 1981.

Al-Anon's Twelve Steps and Twelve Traditions. Revised edition. Virginia Beach: AFG Headquarters Inc., 2005.

Alateen-- a day at a time. Virginia Beach: AFG Headquarters, Inc., 1983.

Alateen--Hope for Children of Alcoholics. Virginia Beach: AFG Headquarters, Inc., 1973.

As We Understood . . . A Collection of Spiritual Insights by Al-Anon and Alateen Members. Virginia Beach: AFG Headquarters, Inc., 1985.

Blueprint for Progress: Al-Anon's Fourth Step Inventory. Virginia Beach: AFG Headquarters, Inc., 2004.

Courage to Be Me: Living with Alcoholism. Virginia Beach: AFG Headquarters, Inc., 1996.

Courage to Change: One Day at a Time in Al-Anon II. Virginia Beach, VA: AFG Headquarters, Inc., 1992.

Discovering Choices: Recovery in Relationships. Virginia Beach: AFG Headquarters, Inc., 2008.

First Steps: Al-Anon . . . 35 Years of Beginnings. Virginia Beach: AFG Headquarters, Inc., 1986.

The Dilemma of the Alcoholic Marriage, Virginia Beach: AFG Headquarters, Inc., 1971.

The Forum: International Monthly Journal of Al-Anon. Virginia Beach: AFG Headquarters, Inc.

From Survival to Recovery: Growing Up in an Alcoholic Home. Virginia Beach: AFG Headquarters, Inc., 1994.

Having Had a Spiritual Awakening . . . Virginia Beach: AFG Headquarters, Inc., 1998.

Homeward Bound. Virginia Beach: AFG Headquarters, Inc., 1993.

Hope for Today. Virginia Beach: AFG Headquarters, Inc., 2002.

How Al-Anon Works for Families and Friends of Alcoholics. Virginia Beach: AFG Headquarters, Inc., 1995.

In All Our Affairs: Making Crises Work for You. Virginia Beach: AFG Headquarters, Inc., 1990.

Living Today in Alateen. Virginia Beach: AFG Headquarters, Inc., 2001.

Living with an Alcoholic. Virginia Beach: AFG Headquarters, Inc., 1966.

Living with Sobriety: Another Beginning. Virginia Beach: AFG Headquarters, Inc., 1979.

Lois Remembers. Virginia Beach: AFG Headquarters, Inc., 1979.

Lois W. and the Pioneers. Two audiotapes. Virginia Beach: AFG Headquarters, Inc., 1982.

Many Voices, One Journey. Virginia Beach: AFG Headquarters, Inc., 2011.

One Day at a Time in Al-Anon. Cornwall, New York: The Cornwall Press, Inc., 1972.

One Day at a Time in Al-Anon. Virginia Beach: AFG Headquarters, Inc., 1973.

On Evolution

Opening Our Hearts, Transforming Our Losses. Virginia Beach: AFG Headquarters, Inc., 2007.

Paths to Recovery: Al-Anon's Steps, Traditions, and Concepts. Virginia Beach: AFG Headquarters, Inc., 1997.

A Pebble in the Pond: The Twelfth Step in Action. Revised 1990. Virginia Beach: AFG Headquarters, Inc., 1976.

Reaching for Personal Freedom: Living the Legacies. Virginia Beach: AFG Headquarters, Inc., 2013.

When I Got Busy, I Got Better. Virginia Beach: AFG Headquarters, Inc., 1994.

GENERAL

Abrams, Nancy Ellen, *A God That Could Be Real: Spirituality, Science, and the Future of Our Planet*. Boston: Beacon Press, 2015.

Adrienne, Carol. *Find Your Purpose, Change Your Life.* New York: HarperCollins, 1999.

Alexander, Bruce. *The Globalization of Addiction: A Study in Poverty of the Spirit.* New York: Oxford University Press, 2008.

Allen, Thomas, *Webster's Ninth New Collegiate Dictionary.* Springfield, Mass: Merriam-Webster Inc., 1983.

Allen, Paula Gunn. *The Sacred Hoop: Recovering the Feminine in American Indian Traditions.* Boston: Beacon Press, 1986.

Amnesty International. *Getting Back on the "Rights Track": A Human Rights Agenda for Canada.* Ottawa, March 2011.

Andrews, Ted. *Animal-Speak: The Spiritual & Magical Powers of Creatures Great and Small.* St. Paul, MN: Llewellyn Publications, 1998.

———. *Animal-Wise: The Spirit Language and Signs of Nature.* Jackson, TN: Dragonhawk Publishing, 2001.

Anonymous. *The Little Red Book: An Interpretation of the Twelve Steps of the Alcoholics Anonymous Program.* Center City, MN: Hazelden Publishing, 1967.

Armstrong, Karen. *Twelve Steps to a Compassionate Life.* Toronto: Alfred A. Knopf Canada, 2010.

Aronson, Elliot. *The Social Animal.* 2nd edition. San Francisco: W.H. Freeman and Company, 1976.

Asher, Ramona. *Women with Alcoholic Husbands.* Chapel Hill: University of North Carolina Press, 1992.

Austin, Gregory. *Alcohol in Western Society from Antiquity to 1800: A Chronological History.* Santa Barbara, CA: ABC-CLIO Information Services, 1985.

Avrich, Paul. *Anarchist Portraits.* Princeton: Princeton University Press, 1988.

Bailey, Joseph. *The Serenity Principle: Finding Inner Peace in Recovery.* San Francisco: Harper, 1990.

Barclay, Harold. *People without Government.* London, U.K.: Kahn and Averill, 1990.

Bartholomew, John C. *The World Atlas.* Edinburgh: John Bartholomew and Son, 1977.

On Evolution

Bateson, Gregory. *Steps to an Ecology of Mind: Collected Essays in Anthropology, Psychiatry, Evolution, and Epistemology.* San Francisco: Chandler Publishing Co., 1972.

Bateson, Gregory et al, *Angels Fear--Toward an Epistemology of the Sacred.* New York: Macmillan, 1987.

Benson, E. F. *Sir Francis Drake.* Edinburgh: U Press, T. A and Constable Ltd., 1927.

Benyus, Janine, *Biomimicry: Innovation Inspired by Nature.* New York: William Morrow and Company Inc., 1997.

Black, Claudia. *It Will Never Happen to Me!* Denver: M. A. C., 1981.

Boger, Robert et al, eds., *Child Nurturance: Volume 4, Child Nurturing in the 1980s.* New York: Plenum Press, 1984.

Bookchin, Murray. *The Ecology of Freedom: The Emergence and Dissolution of Hierarchy.* Montreal: Black Rose Books, 1991.

———. *The Philosophy of Social Ecology.* Montreal: Black Rose Books, 1990.

———. *Remaking Society.* Montreal: Black Rose Books, 1989.

Borchert, William D. *The Lois Wilson Story: When Love Is Not Enough.* Center City, MN: Hazelden, 2005.

Borkman, T. *Understanding Self-Help and Mutual Aid: Experiential Learning in the Commons.* New Brunswick, NJ: Rutgers University Press. 1999

Bornstein, David, *How to Change the World: Social Entrepreneurs and the Power of New Ideas.* New York: Oxford University Press, Inc., 2004

Boucher, Douglas, ed., *The Biology of Mutualism: Ecology & Evolution*. London, U.K.: Crown Helm, 1985.

Bowlby, John. *Charles Darwin: A Biography*. London: Hutchinson, 1990.

Bowles, Samuel et al, *A Cooperative Species: Human Reciprocity and its Evolution*. Princeton, NJ : Princeton University Press, 2011.

Bradshaw, John. *Healing the Shame that Binds You*. Deerfield Beach, FL.: Health Communications Inc., 1988.

Brand, Russell. *Recovery: Freedom from Our Addictions*, London, U.K.: Pan Macmillan, 2017.

Brandt, Willy et al, *North-South: A Programme for Survival*. London, U.K.: Pan Books Ltd., 1980.

Breton, Denise et al, *The Paradigm Conspiracy: Why Our Social Systems Violate Human Potential--And How We Can Change Them*. Center City, MN: Hazelden, 1996.

Bridges, E. Lucas. *Uttermost Part of the Earth*. London, U.K.: Hodder and Stoughton, 1948.

Brock, Thomas. *A Eutrophic Lake: Lake Mendota, Wisconsin*, New York: Springer- Verlag, 1985.

Brown, Dee. *Bury My Heart at Wounded Knee: An Indian History of the American West*. New York: Henry Holt and Company, Thirtieth Anniversary edition, 2001.

Brown, Lester. *World on the Edge: How to Prevent Environmental and Economic Collapse*. Earth Policy Institute. New York: W.W. Norton & Company, 2011.

On Evolution

Brown, Stephanie et al, *The Alcoholic Family in Recovery*. New York: The Guilford Press, 1999.

Browne, Janet. *Charles Darwin: Voyaging. A Biography*. New York: A. Knopf, 1995.

Browne-Miller, Angela. *The Praeger International Collection on Addictions. Volume 1: Faces of Addiction, Then and Now*. Westport, CT: Praeger, 2009.

Brundtland, Gro Harlem et al, *Our Common Future*. The World Commission on Environment and Development. New York: Oxford University Press, 1987.

Bufe, Charles. *Alcoholics Anonymous: Cult or Cure?* San Francisco: Sharp Press, 1991.

Burkhardt, Frederick et al, eds., *A Pluralistic Universe: William James*. Cambridge, MA: Harvard University Press, 1977.

Butler, Katy. "Spirituality and Therapy." *Utne Reader*, no. 43 (Jan./Feb. 1991).

Campbell, Joseph. *The Masks of God: Primitive Mythology*. New York: Penguin Books, 1969.

———. *The World of Joseph Campbell: Transformations of Myth through Time*. VHS tapes. Introduction: The Hero's Journey. 57 minutes, 1987.

Campfens, Hubert, ed., *Community Development around the World: Practice, Theory, Research, Training*. Toronto: University of Toronto Press, 1997.

Capra, Fritjof et al, *Green Politics*. New York: E. P. Dutton Inc., 1984.

Carroll, Joseph, ed., *On "The Origin of Species": Charles Darwin*. Orchard Park, NY: Broadview, 2003.

Carson, Rachel. *Silent Spring*. Boston: Houghton Mifflin, 1962.

Cathcart, Thomas et al, *Plato and a Platypus Walk into a Bar...*New York: Penguin Books, 2007, 29.

Cavanagh, John et al, eds., *Alternatives to Economic Globalization: A Better World is Possible*. San Francisco: Berrett-Koehler Publishers, 2004.

Chapman, Anne. *Drama and Power in a Hunting Society: The Selk'nam of Tierra del Fuego*. New York: Cambridge University Press, 1982.

Cheever, Susan. *My Name Is Bill*. New York: Washington Square Press, 2004.

Chesnut, Glenn F., *Father Ralph Pfau and the Golden Books: The Path to Recovery from Alcoholism and Drug Addiction*. Hindsfoot Foundation, Bloomington, IN: iUniverse, 2017.

Chopra, Deepak. *Overcoming Addictions: The Spiritual Solution*. New York: Harmony Books, 1997.

Clancy, Kelly. "Survival of the Friendliest" in New York: Nautilus, Issue 046, March 23, 2017.

Cleveland, Martha. *Chronic Illness and the Twelve Steps: A Practical Approach to Spiritual Resilience*. Center City, MN: Hazelden, 1999.

Co-dependents Anonymous. *Co-dependents Anonymous*. Phoenix, AZ: Co-dependents Anonymous, 1995.

On Evolution

Coelho, Paulo. *The Pilgrimage: A Contemporary Quest for Ancient Wisdom*. Translated by Alan Clarke. New York: HarperCollins Publishers, 1995.

Covington, Stephanie. *A Woman's Way through the Twelve Steps*. Center City, MN: Hazelden, 1994.

Coyhis, Don. *The Red Road to Wellbriety--in the Native American Way*. Aurora, CO: Coyhis Publishing, Inc., 2002.

———. *The Wellbriety Movement Comes of Age: The Fulfillment of Prophecy*. Aurora, CO: Coyhis Publishing, Inc., 2011.

Coyhis, Don et al, *Alcohol Problems in Native America: The Untold Story of Resistance and Recovery--"The Truth about the Lie."* Aurora, CO: Coyhis Publishing, Inc., 2006.

Darwin, Charles. *The Autobiography of Charles Darwin*. Edited by Nora Barlow. London, U.K.: Collins, 1958.

———. *The Descent of Man*. Chicago: Rand, McNally and Company, revised edition 1874. Originally published 1871.

———. *The Origin of Species*. Harmondsworth, U.K.: Penguin Books, 1st edition, 1968, originally published 1859.

———. *The Origin of Species*. Everyman's Library. New York: Dutton, 1872.

———. *The Voyage of the Beagle*. Hertfordshire, U.K.: Wordsworth Editions Limited, 1997, first published 1845.

Daschuk, James. *Clearing the Plains: Disease, Politics of Starvation, and the Loss of Aboriginal Life*. Regina, SK: University of Regina Press, 2013.

Dass, Ram et al, *Compassion in Action: Setting Out on the Path of Service*. New York: Bell Tower, 1992.

Dayton, Tian. *Trauma and Addiction*. Deerfield Beach, FL: Health Communications, Inc., 2000.

Dennett, Daniel. *Darwin's Dangerous Idea: Evolution and the Meanings of Life*. New York: Simon & Schuster, 1995.

Denzin, Norman et al, *The Recovering Alcoholic*. Newbury Park, CA: Sage Publications, Inc., 1987.

De Paolo, Charles. *The Ethnography of Charles Darwin: A Study of His Writings on Aboriginal Peoples*. Jefferson, NC: McFarland & Co., Inc., 2010.

Department of Justice Canada. *A Consolidation of the Constitution Acts 1867 to 1982*. Ottawa, ON: April 17, 1982.

Desmond, Adrien, et al, *Darwin*. London, U.K.: Michael Joseph, 1991.

———. *Darwin's Sacred Cause: Race, Slavery and the Quest for Human Origins*. London, U.K.: Allen Lane, 2009.

Diamond, Jared. *Collapse: How Societies Choose to Fail or Succeed*. New York: Viking Press, 2005.

———. *Guns, Germs, and Steel: The Fates of Human Societies*. New York: W.W. Norton & Company, Inc., 1999.

D'Orbigny, Alcide. *Voyages dans des deux Amériques*. Paris: Furne, Jouvet and Co., 1857.

Dossey, Larry. *Recovering the Soul: A Scientific and Spiritual Search*. New York: Bantam Books, 1989.

On Evolution

Doyle, Dr. Robert et al, *Almost Alcoholic: Is My (or My Loved One's) Drinking a Problem?* Center City, MN: Hazelden Publishing, 2012.

Dubos, Rene. *Mirage of Health: Utopias, Progress, and Biological Change.* New Brunswick, NJ: Rutgers University Press, 1959.

Duncan, James. "The Low Bottom Family Disease and Early Pioneering in the Twelve Step Movement." Manuscript, 2005.

_____*Living the Twelve Steps: Change Ourselves and Change the World.* Gatineau, QC: BAICO Publishing, 2005.

_____*World Awakening: A.A. and Higher Power Mutual Aid.* Renfrew ON: General Store Publishing House, 2011.

Dupont, Robert et al, *A Bridge to Recovery: An Introduction to Twelve-Step Programs.* Washington, D.C.: American Psychiatric Press, Inc., 1994.

Durant, John, ed. *Darwinism and Divinity: Essays on Evolution and Religious Belief.* New York: Basil Blackwell, Inc., 1985.

The Ecologist. *A Blueprint for Survival.* Harmondsworth, U.K.: Penguin Books, 1972.

Ehrlich, Paul. *The Machinery of Nature.* New York: Simon & Schuster, 1986.

Ehrlich, Paul et al, *Healing the Planet.* New York: Addison–Wesley Publishing Co., Inc., 1991.

Ehrlich, Paul et al, *New World New Mind.* New York: Doubleday, 1989.

Eisler, Riane. *The Chalice and the Blade: Our History, Our Future.* New York: Harper and Row, 1987.

Elk, Black, *Black Elk Speaks: Being the Life Story of a Holy Man of the Oglala Sioux*. As told through John G. Neihardt, Lincoln, NE: University of Nebraska Press, 1979, originally published 1932.

Evans, K. et al, *Treating Addicted Survivors of Trauma*. New York: Guilford, 1995.

Fainzang, Sylvie. "When Alcoholics Are Not Anonymous." *Medical Anthropology Quarterly* 8, no. 3 (1994): 336–45.

Feest, Christian, ed. *Indians and Europe: An Interdisciplinary Collection of Essays*. Lincoln, NE: University of Nebraska Press, 1989.

Ferguson, Marilyn. *The Aquarian Conspiracy: Personal and Social Transformation in Our Time*. Los Angeles: J.P. Tarcher, Inc., 1980.

Fisher, Andy. *Radical Ecopsychology: Psychology in the Service of Life*. Albany, NY: State University of New York Press, 2002.

Fishman, Ross. *Alcohol and Alcoholism*. New York: Chelsea House Publishers, 1987.

Flourny, Th. *The Philosophy of William James*. Authorized translation by Edwin Holt et al, Freeport, NY: Books for Libraries Press, 1969. Originally published 1917.

Fossey, Dian. *Gorillas in the Mist*. Boston: Houghton Mifflin, 1983.

Fuller, R. Buckminster. *Operating Manual for Spaceship Earth*. New York: Pocket Books, 1970.

Fuller, Robert. *Spiritual but Not Religious: Understanding Unchurched America*. New York: Oxford University Press, 2001.

Gaer, Joseph. *What the Great Religions Believe*. New York: New American Library, 1963.

Gelman, Irving. *The Sober Alcoholic: An Organizational Analysis of Alcoholics Anonymous*. New Haven, CT: College and University Press, 1964.

George-Kanentiio, Doug. *Iroquois Culture and Commentary*. Santa Fe, NM: Clear Light Publishers, 2000.

Gilliam, Marianne. *How Alcoholics Anonymous Failed Me*. New York: Eagle Brook, 1998.

Gitterman, Alex et al, eds., *Mutual Aid Groups, Vulnerable Populations, and the Life Cycle*. 2nd edition. New York: Columbia University Press, 1994.

Glassner, Barry. *The Culture of Fear*. New York: Basic Books, 1999.

Golding, P., ed. *Alcoholism: Analysis of a World-wide Problem*. Hingham, MA: MTP Press Ltd., 1983.

Good Tracks, J.G. "Native American Non-interference." *Social Work*, November 1973, 30–34.

Grant, John et al, *Why Can't I Stop?* Baltimore: Johns Hopkins University Press, 2016.

Graveline, F.J. *Circle Works: Transforming Eurocentric Consciousness*. Halifax, NS: Fernwood Publishing, 1998.

Gravitz, Herbert et al, *Recovery: A Guide for Adult Children of Alcoholics*. New York: Simon & Schuster, 1985.

Green, Rayna et al, *The Encyclopedia of the First Peoples of North America.* Toronto: Douglas & McIntyre, 1999.

Greenleaf, Jael. "Co-alcoholic-Para-Alcoholic: Who's Who and What's the Difference?" Los Angeles: The Alcoholism Center for Women, 1981. Paper presented at the National Council on Alcoholism, 1982 Annual Alcoholism Forum, New Orleans.

Green Party of Canada. "The Green Party of Canada Campaign 1993." Ottawa, ON, 1993.

Green Party of Canada. "Vision Green." GPC Policy Document. Ottawa, ON, April 2011.

Greg, Richard, *The Value of Voluntary Simplicity.* Wallingford, PA: Pendle Hill, 1936.

Guenther, Heinz. *The Footprints of Jesus' Twelve in Early Christian Traditions.* New York: Peter Lang, 1985.

Gusinde, Martin. *The Yamana: The Life and Thought of the Water Nomads of Cape Horn.* Translated from the German by Frieda Schutze. New Haven, CT: Human Relations Area Files, 1961.

Hamilton, Tim. *The 12 Steps and Dual Disorders: A Framework for Recovery for Those of Us with Addiction and an Emotional or Psychiatric Illness.* Center City, MN: Hazelden, 1994.

Hammer, Leon. *dragon rises, red bird flies: Psychology and Chinese Medicine.* Barrytown, NY: Station Hill Press, Inc., 1990.

Hammond, William, ed. *12 Step Wisdom at Work: Transforming Your Life and Your Organization.* Dover, NH: Kogan Page U.S., 2001.

On Evolution

Hampden, John, ed. *Francis Drake, Privateer*. London, U.K.: Eyre Methuen, 1972.

Hardin, Garret. "The Tragedy of the Commons." *Science: Journal of the American Association for the Advancement of Science* 162, no. 3859 (Dec. 13, 1968).

Harman, Jay. *The Shark's Paintbrush: Biomimicry and How Nature Is Inspiring Innovation*. Ashland, OR: White Cloud Press, 2013.

Harpur, Tom. *Living Waters: Selected Writings on Spirituality*. Toronto: Thomas Allen Publishers, 2006.

Harries-Jones, Peter. *A Recursive Vision: Ecological Understanding and Gregory Bateson*. Toronto: University of Toronto Press, 1995.

Hartigan, Francis. *Bill W.: A Biography of Alcoholics Anonymous Cofounder Bill Wilson*. New York: St. Martin's Press, 2000.

Hawken, Paul. *The Ecology of Commerce: A Declaration of Sustainability*. New York: HarperCollins, 1993.

Hawken, Paul et al, *Natural Capitalism: Creating the Next Industrial Revolution*. New York: Little, Brown and Company, 1999.

Hazlewood, Nick. *Savage: The Life and Times of Jemmy Button*. London, U.K.: Hodder and Stoughton, 2000.

Herman, Judith. *Trauma and Recovery*. New York: Basic Books, 1992.

Heyman, Gene M. *Addiction: A Disorder of Choice*. Cambridge, MA: Harvard University Press, 2009.

Himmelfarb, Gertrude. *Darwin and the Darwinian Revolution*. New York: W.W. Norton & Company Inc., 1962. First published 1959.

Hirschfield, Jerry. *The Twelve Steps for Everyone . . . Who Really Wants Them.* Rev. ed. Center City, MN: Hazelden Publishing, 1990.

Hornbacher, Marya. *Sane: Mental Health, Addiction, and the Twelve Steps.* Center City, MN: Hazelden Publishing, 2010.

Howard, Ted et al, *Who Should Play God? The Artificial Creation of Life and What It Means for the Future of the Human Race.* New York: Dell Publishing Co., Inc., 1977.

Humphreys, Keith. *Circles of Recovery: Self-Help Organizations for Addictions.* New York: Cambridge University Press, 2004.

Hunter, Charlotte et al, *Women Pioneers in Twelve Step Recovery.* Center City, MN: Hazelden, 1999.

Huxley, Aldous. *Brave New World.* New York: Perennial Library, 1988. Originally published 1932.

Jacobson, Bertil et al, "Obstetric Care and Proneness of Offspring to Suicide as Adults: A Case-Control Study." *British Medical Journal* 317, no. 1346 (Nov. 14, 1998).

James, William. *The Varieties of Religious Experience.* New York: Mentor Books, 1958. Originally published 1902.

Jastrow, Robert. *God and the Astronomers.* Second edition. New York: W.W. Norton & Company, Inc. 2000.

_____General Editor, *The Essential Darwin*, Toronto: Little, Brown and Company (Canada) Ltd., 1984.

Janzen, John. "Drums Anonymous: Towards an Understanding of Structures of Therapeutic Maintenance." In M. W. de Vries et al, eds., *The Use and Abuse of Medicine*, 154–66. New York: Praeger, 1982.

Joyner, Tim. *Magellan*. Camden, ME: International Marine, 1992.

Kaminer, Wendy. *I'm Dysfunctional, You're Dysfunctional: The Recovery Movement and Other Self-Help Fashions*. Reading, MA: Addison–Wesley Publishing Company, Inc., 1992.

Kasl, Charlotte. *Many Roads, One Journey: Moving beyond the Twelve Steps*. New York: Harper Perennial, 1992.

Katz, Alfred et al, writers and editors. *The Strength in Us: Self-help Groups in the Modern World*. New York: Franklin Watts, 1976.

Katz, Alfred. *Self-Help in America: A Social Movement Perspective*. New York: Twain Publishers, 1993.

Kawanabe, Hiroya et al, eds., *Mutualism and Community Organization: Behavioral, Theoretical, and Food-Web Approaches*. Oxford: Oxford University Press, 1993.

Keating, Thomas. *Divine Therapy and Addiction: Centering Prayer and the Twelve Steps*. New York: Lantern Books, 2009.

Kelsey, H. *Sir Francis Drake: The Queen's Pirate*. New Haven, CT: Yale University Press, 1998.

Khanna, Parag. *How to Run the World: Charting a Course to the Next Renaissance*. New York: Random House, 2011.

Kingwell, Mark. *The World We Want: Virtue, Vice, and the Good Citizen*. Toronto: Viking, 2000.

Kitchell, James, ed., *Food Web Management: A Case Study of Lake Mendota.* New York: Springer-Verlag, 1992.

Kriesberg, Louis et al, eds., *Research in Social Movements, Conflicts and Change: A Research Annual.* Greenwich, CT: JAI Press Inc., 1978.

Kropotkin, Peter. "Communism and Anarchy." In *Small Communal Experiments and Why They Fail.* Petersham, Australia: Jura Books, 1901, republished 1997.

———. *Ethics: Origin and Development.* Montreal: Black Rose Books, 1992. First published posthumously, 1924.

———. *Fields, Factories and Workshops.* Montreal: Black Rose Books, 1994. First published 1898.

———. *Memoirs of a Revolutionist.* New York: Grove Press, Inc., 1968. First published 1899.

———. *Mutual Aid: A Factor of Evolution.* Montreal: Black Rose Books, 1902. Republished 1989.

———. "Proposed Communist Settlement: A New Colony for Tyneside or Wearside." In *Small Communal Experiments and Why They Fail.* Petersham, Australia: Jura Books, 1997, first published 1895.

Kulchyski, Peter et al, eds., *In the Words of Elders: Aboriginal Cultures in Transition.* Toronto: University of Toronto Press, 2003. First published 1999.

Kurtz, Ernest. *A.A.: The Story.* San Francisco: Harper and Row, 1988. Revised edition of *Not-God: A History of Alcoholics Anonymous,* 1979.

———. *Not God: A History of Alcoholics Anonymous.* Center City, MN: Hazelden Educational Services, 1979.

On Evolution

Kurtz, Ernest et al, *The Spirituality of Imperfection*. New York: Bantam Books, 1992.

Lachance, Albert. *Cultural Addiction: The Greenspirit Guide to Recovery*. Berkeley, CA: North Atlantic Books, 2006.

Larsen, Earnie. *Stage II Recovery: Life beyond Addiction*. San Francisco: Harper and Row, 1985.

Lavoie, Francine et al, eds., *Self Help and Mutual Aid Groups: International and Multicultural Perspectives*. New York: The Haworth Press, Inc., 1994.

Lawson, Anne et al, *Alcoholism and the Family: A Guide to Treatment and Prevention*. 2nd ed. Gaithersberg, MD: Aspen Publishers, Inc., 1998.

Lee, Dora. *Biomimicry: Inventions Inspired by Nature*. Toronto: Kids Can Press Ltd., 2011.

Leonard, L. S. *Witness to the Fire: Creativity and the Veil of Addiction*. Boston: Shambhala Publications, 1989.

Lestringant, Frank. "The Myth of the Indian Monarchy: An Aspect of the Controversy between Thevet and Lery." In Christian Feest, ed., *Indians and Europe: An Interdisciplinary Collection of Essays*. Lincoln, NE: University of Nebraska Press, 1989.

Levinson, Henry. *The Religious Investigations of William James*. Chapel Hill, NC: University of North Carolina Press, 1981.

Lieberman, Joshua, ed., *New Trends in Group Work*. New York: Association Press, 1939.

Lobdell, Jared. *This Strange Illness: Alcoholism and Bill W.* New York: Aldine de Gruyter, 2004.

Lofland, John. *Analyzing Social Settings: A Guide to Qualitative Observation and Analysis*. Belmont, CA: Wadsworth Publishing Company, Inc., 1971.

Lovelock, James. *The Ages of Gaia: A Biography of Our Living Earth*. New York: Bantam Books, 1990.

———. *Gaia: A New Look at Life on Earth*. London, U.K.: Oxford University Press, 1979.

———. *The Revenge of Gaia*. London, U.K.: Allen Lane, Penguin Books, 2006.

Loye, David, ed., *The Evolutionary Outrider: The Impact of the Human Agent on Evolution*. Westport, CT: Praeger, 1998.

Makela, Klaus et al, *Alcoholics Anonymous as a Mutual-Help Movement: A Study in Eight Societies*. Madison, WI: University of Wisconsin Press, 1996.

Mander, Jerry et al, eds., *Paradigm Wars: Indigenous Peoples' Resistance to Globalization*. San Francisco: Sierra Club Books, 2006.

Maracle, Brian. *Crazywater: Native Voices on Addiction and Recovery*. Toronto: Viking, 1993.

Marean, Curtis. "The Most Invasive Species of All." *Scientific American*, August 2015, Vol. 313, No. 2.

Margulis, Lynn et al, eds., *Symbiosis as a Source of Evolutionary Innovation*. Cambridge, MA: The MIT Press, 1991.

Marshall III, Joseph. *The Journey of Crazy Horse: A Lakota History*. New York: Viking, the Penguin Group. 2004.

On Evolution

Massie, Joseph et al, *Managing: A Contemporary Introduction*. 3rd edition. Englewood Cliffs, NJ: Prentice–Hall, Inc., 1981.

Matthiessen, Peter. *In the Spirit of Crazy Horse*. New York: Viking Penguin, 1991, first published 1983.

Maxwell, Milton. *The Alcoholics Anonymous Experience*. New York: McGraw Hill, 1984.

McCalman, Iain. *Darwin's Armada*. London, UK: Simon & Schuster, 2009.

McGovern, Patrick. *Uncorking the Past: The Quest for Wine, Beer, and Other Alcoholic Beverages*. Berkeley, CA: University of California Press, 2009.

McKew Parr, C. *Ferdinand Magellan, Circumnavigator*. New York: Thomas Y. Crowell, 1964.

McKnight, John. *The Careless Society: Community and Its Counterfeits*. New York: Basic Books, 1995.

Mead, Margaret, ed. *Cooperation and Competition among Primitive Peoples*. Boston: Beacon Press, 1961. First published 1937.

Meadows, Dennis et al, *The Limits to Growth*. The Club of Rome. New York: Universe Books, 1972.

Means, Russell. *Where White Men Fear to Tread: The Autobiography of Russell Means*. New York: St. Martin's Griffin, 1995.

Menaker, Esther. *Otto Rank: A Rediscovered Legacy*. New York: Columbia University Press, 1982.

Miller, Martin. *Kropotkin*. Chicago: The University of Chicago Press, 1976.

Miller, William et al, eds., *Treating Addictive Behaviors*. New York: Plenum Press, 1986.

Mitroff, Ian et al, *Framebreak: The Radical Redesign of American Business*. San Francisco: Jossey Bass Publishers, 1994.

Montagu, Ashley. *Darwin: Competition and Cooperation*. New York: Henry Schuman, 1952.

Morris, Brian. *Kropotkin: The Politics of Community*. Amherst, NY: Humanity Books, 2004.

Morris, Gwen. *Gwen: Memories and Sketches from the Life of Gwen Morris*, editor John Morris, Alexandria, ON: The Printing House, 2014.

Mowrer, O.H. *The Crisis in Psychiatry and Religion*. Princeton, NJ: Van Nostrand, 1961.

———. "Peer Groups and Medication: The Best 'Therapy' for Laymen and Professionals Alike." *Psychotherapy: Theory, Research and Practice*. Vol. 8, no. 1 (1971): 44–54.

———. "The 'Self-Help' or Mutual Aid Movement: Do Professionals Help or Hinder?" In *Self Help and Health: A Report*. New York: New Human Sciences Institute, 1976.

Mumford, Lewis. *The Culture of Cities*. New York: Harcourt, Brace and Company, 1938.

Murphy, Brian K. *Transforming Ourselves, Transforming the World: An Open Conspiracy for Social Change*. New York: Zed Books, 1999.

Nabokov, Peter. *Native American Testimony: A Chronicle of Indian–White Relations from Prophecy to the Present, 1492–1992*. New York: Penguin Books, 1991.

On Evolution

National Film Board of Canada. *Goddess Remembered*. Studio D, 54-minute videotape, Ottawa, 1989.

Neuman, W. L. *Social Research Methods: Qualitative and Quantitative Approaches*. Boston: Allyn & Bacon, 1994.

Newburg, Andrew et al, *Why God Won't Go Away: Brain Science and the Biology of Belief*. New York: Ballantine Books, 2001.

Novak, Michael, ed. *Democracy and Mediating Structures: A Theological Inquiry*. Washington, D.C.: American Institute for Public Policy Research, 1980.

Nowak, Martin et al, *SuperCooperators: Altruism, Evolution, and Why We Need Each Other to Succeed*. New York, NY: Free Press, 2011.

Nowinski, Joseph. *If You Work It, It Works! The Science Behind 12 Step Recovery*. Center City, MN: Hazelden Publishing, 2015.

O'Byrne, Darren. *The Dimensions of Global Citizenship: Political Identity beyond the Nation-State*. London, U.K.: Frank Cass and Co. Ltd., 2003.

Oliver-Diaz, Philip et al, *Twelve Steps to Self-Parenting*. Deerfield Beach, FL: Health Communications, Inc., 1988.

O'Murchu, Diarmuid. *Religion in Exile: A Spiritual Homecoming*. New York: The Crossroad Publishing Company, 2000.

Osofsky, Stephen. *Peter Kropotkin*. Boston, MA: Twayne Publishers, a Division of G. K. Hall & Company, 1979.

Palmer, Trevor. *Perilous Planet Earth: Catastrophe and Catastrophism through the Ages*. Cambridge, MA: Cambridge University Press, 2003.

Paton, Alan. *Cry, The Beloved Country.* New York: Scribner, 2003. Originally published 1948.

Patton, M. Q. *Qualitative Evaluation and Research Methods.* 2nd ed. Newbury Park: Sage Publications, 1990.

Peck, M. Scott. *The Different Drum: Community Making and Peace.* New York: Simon & Schuster, 1987.

———. *Further along the Road Less Traveled. Addiction: The Sacred Disease.* Audiotape. New York: Simon & Schuster, 1991.

———. *The Road Less Traveled.* New York: Simon & Schuster, 1978.

Peckham, Morse. *"The Origin of Species" by Charles Darwin: A Varorium Text.* Philadelphia: University of Philadelphia Press, 1959.

Pettegree, Andrew, ed., *The Early Reformation in Europe.* Cambridge, MA: Cambridge University Press, 1992.

Pittman, David et al, eds., *Society, Culture, and Drinking Patterns.* New York: John Wiley & Sons, Inc., 1962.

Plaskow, Judith et al, *Weaving the Visions: New Patterns in Feminist Spirituality.* New York: HarperCollins Publishers, 1989.

Powell, Thomas. *Self-Help Organizations and Professional Practice.* Silver Spring, MD: National Association of Social Workers, 1987.

Powell, Thomas, ed. *Understanding the Self-Help Organization: Framework and Findings.* Thousand Oaks, CA: Sage Publications, 1994.

Quammen, David. *The Reluctant Mr. Darwin: An Intimate Portrait of Charles Darwin and the Making of His Theory of Evolution.* New York: W.W. Norton & Company, 2006.

Rank, Otto. *The Trauma of Birth.* New York: Harper and Row, Publishers, 1973. First edition 1929.

Raphael, Matthew J. *Bill W. and Mr. Wilson: The Legend and Life of A.A.'s Cofounder.* Amherst: University of Massachusetts Press, 2000.

Rapping, Elaine. *The Culture of Recovery: Making Sense of the Self-Help Movement in Women's Lives.* Boston: Beacon Press, 1996.

Ridley, Matt. *The Agile Gene: How Nature Turns On Nurture.* Toronto : Harper*Perennial*Canada, 2003.

_____. *The Origins of Virtue: Human Instincts and the Evolution of Cooperation.* New York: Viking, 1996.

Rifkin, Jeremy. *The Empathic Civilization: The Race to Global Consciousness in a World of Crisis.* New York: Tarcher, 2009.

Roberts, Monty. *Horse Sense for People.* Toronto: Alfred A. Knopf Canada, 2001.

Robertson, Nan. *Getting Better: Inside Alcoholics Anonymous.* New York: William Morrow and Co., 1988.

Robinson, David. *Talking Out of Alcoholism: The Self-Help Process of Alcoholics Anonymous.* London, U.K.: Crown Helm, 1978.

Romeder, Jean–Marie et al, *The Self Help Way: Mutual Aid and Health.* Ottawa: Canadian Council on Social Development, 1990.

Room, Robin. "Alcoholics Anonymous as a Social Movement." In Barbara McCrady et al, eds., *Research on Alcoholics Anonymous: Opportunities and Alternatives*, 167–87. New Brunswick, NJ: Rutgers Center of Alcohol Studies, 1993.

———. "Healing Ourselves and Our Planet: The Emergence and Nature of a Generalized Twelve-Step Consciousness." *Contemporary Drug Problems*, Vol. 19, no. 4 (1992): 717–40.

Rotgers, Frederick et al, eds., *Treating Substance Abuse*. New York: The Guilford Press, 1996.

Rousseau, Jean-Jacques. *A Discourse on Inequality*. London: Penguin Books, 1984, first published 1755.

Ruiz, Don Miguel. *The Four Agreements: A Toltec Wisdom Book*. San Rafael, CA: Amber–Allen Publishing, 1997.

Runkle, Gerald. *Anarchism: Old and New*. New York: Dell Publishing Co., Inc., 1972.

Ryley, Nancy. *The Forsaken Garden: Four Conversations on the Deep Meaning of Environmental Illness*. Wheaton, IL: The Theosophical Publishing House, 1998.

Sabini, Meredith, ed., *The Earth Has a Soul: C.G. Jung on Nature, Technology & Modern Life*. Berkeley, CA: North Atlantic Books, 3rd edition, 2008.

Saloman, Frank et al, eds., *The Cambridge History of the Native Peoples of the Americas*. Vol. III, South America, Part 2. Cambridge: Cambridge University Press, 1999.

Sandoz, Mari, *Crazy Horse: Strange Man of the Oglalas*. Lincoln, NE: University of Nebraska Press, 1961.

On Evolution

Sapp, Jan. *Evolution by Association: A History of Symbiosis*. Oxford: Oxford University Press, 1994.

Saul, John Ralston. *On Equilibrium*. Toronto: Viking, 2001.

Sears, Paul. *Charles Darwin: The Naturalist as a Cultural Force*. New York: Charles Scribner's Sons, 1950.

Schaef, Anne Wilson. *Beyond Therapy, Beyond Science: A New Model for Healing the Whole Person*. San Francisco: Harper and Row, 1992.

———. *When Society Becomes an Addict*. San Francisco: Harper and Row, 1987.

———. *The Addictive Organization*. San Francisco: Harper and Row, 1988.

Schell, Jonathan. *The Fate of the Earth*. New York: Alfred A. Knopf, 1982.

Schumacher, E. F. *Small Is Beautiful: Economics as if People Mattered*. Vancouver, BC: Hartley and Marks Publishers Inc., 1973. Revised edition published 1999.

Schwartz, Barry et al, *Practical Wisdom: The Right Way to Do the Right Thing*. New York: Riverhead Books, 2010.

Serrano, Miguel. *C.G. Jung and Hermann Hesse: A Record of Two Friendships*. London, U.K.: Routledge & Kegan Paul, 1966.

Shealy, C. Norman et al, *The Creation of Health: The Emotional, Psychological, and Spiritual Responses that Promote Health and Healing*. Walpole, NH: Stillpoint Publishing, 1993.

Shepard, Paul et al, eds., *The Subversive Science: Essays toward an Ecology of Man*. Boston: Houghton Mifflin, 1969.

Simos, Miriam. *The Earth Path: Grounding Your Spirit in the Rhythms of Nature.* San Francisco: Harper Collins Publishers, 2004.

Singer, Peter. *A Darwinian Left: Politics, Evolution and Cooperation.* New Haven: Yale University Press, 1999.

Sioui, Georges E. *For an Amerindian Autohistory: An Essay on the Foundations of a Social Ethic.* Translated from the French. Montreal: McGill–Queen's University Press, 1992.

———. *Huron–Wendat: The Heritage of the Circle.* Rev. ed. Vancouver: University of British Columbia Press, 1999.

Smith, Bob et al, *Children of the Healer: The Story of Dr. Bob's Kids.* Park Ridge, IL: Parkside Publishing Corporation, 1992.

Smith, Robert Leo, ed., *The Ecology of Man: An Ecosystem Approach.* Second edition. New York: Harper and Row, 1972.

Sonkin, Daniel J. *Wounded Boys, Heroic Men.* Stamford, CT: Longmeadow Press, 1992.

Speerstra, Karen. *The Green Devotional: Active Prayers for a Healthy Planet.* San Francisco: Conari Press, 2010.

Spretnak, Charlene. *The Spiritual Dimension of Green Politics.* Santa Fe, NM: Bear & Company, Inc., 1986.

Standage, Tom. *A History of the World in 6 Glasses.* New York: Walker & Company, 2005.

Stempler, B. L. et al, eds., *Social Group Work Today and Tomorrow.* New York: The Haworth Press, 1996.

On Evolution

Stone, Merlin. *When God Was a Woman*. New York: Harcourt Brace & Company, 1976.

Sullivan, Dennis. *The Mask of Love: Corrections in America, Toward a Mutual Aid Alternative*. Port Washington, NY: Kennicat Press, 1980.

Suzuki, David. *The Sacred Balance: Rediscovering Our Place in Nature*. Toronto: The Douglas and McIntyre Publishing Group, 1997.

Suzuki, David et al, *Good News for a Change: Hope for a Troubled Planet*. Toronto: Stoddart Publishing Company Ltd., 2002.

Thayer, Bradley. *Darwin and International Relations: On the Evolutionary Origins of War and Ethnic Conflict*. Lexington: University Press of Kentucky, 2004.

Thomsen, Robert. *Bill W.* Center City, MN: Hazelden, 1975.

Thomson, Keith. *The Young Charles Darwin*. New Haven, CT: Yale University Press, 2009.

Todes, Daniel P. *Darwin without Malthus: The Struggle for Existence in Russian Evolutionary Thought*. New York: Oxford University Press, 1989.

Tolle, Eckhart. *A New Earth: Awakening to Your Life's Purpose*. New York: Dutton, 2005.

Tolstoy, Leo. *Anna Karenina*. New York: Signet Classic, 1961. First published 1877.

Tomasello, Michael. *Why We Cooperate*. Cambridge, MA: The MIT Press, 2009.

Torrance, Tom. *Dar es Salaam, 1963,* Renfrew, ON: General Store Publishing House, 2010.

Trager, William. *Symbiosis.* New York: Van Nostrand Reinhold Co., 1970.

Trevino, A.J. "Alcoholics Anonymous as Durkheimian Religion." *Research in the Social Scientific Study of Religion.* Vol. 4 (1992): 183–208.

UNICEF. "Child Deaths by Injury in Rich Nations." *Innocenti Report Card* Issue Number 2. Florence, Italy, February 2001.

Vaillant, George. *The Natural History of Alcoholism.* Cambridge, MA: Harvard University Press, 1983.

Valverde, Mariane et al, "One Day at a Time and . . ." *Sociology: The Journal of the British Sociological Association* 33, no. 2 (1999): 393–410.

Van den Burgh, Nan, ed. *Feminist Perspectives on Addictions.* New York: Springer Publishing Company, 1991.

Vanden Burkdt, Robert. *The Religious Philosophy of William James.* Chicago: Nelson–Hall, 1981.

van der Dennen, Johan et al, eds., *The Darwinian Heritage and Sociobiology.* Westport, CT: Praeger, 1999.

van der Kolk, Bessel et al, eds., *Traumatic Stress: The Effects of Overwhelming Experience on Mind, Body, and Society.* New York: The Guilford Press, 1996.

Wackernagel, Mathis et al, *Our Ecological Footprint: Reducing Human Impact on the Earth.* Gabriola Island, BC: New Society Publishers, 1996.

Wallis, Claudia. "Faith and Healing." *Time,* June 24, 1996.

On Evolution

Walsch, Neale Donald. *Conversations with God: An Uncommon Dialogue, Book 1*. New York: G. P. Putnam's Sons, 1995.

Ward, Barbara et al, *Only One Earth: The Care and Maintenance of a Small Planet*. New York: W.W. Norton and Company, Inc., 1972.

Wasserman, Harry et al, *The Human Bond: Support Groups and Mutual Aid*. New York: Springer Publishing Company, 1988.

Weatherford, Jack. *Savages and Civilization: Who Will Survive?* New York: Crown Publishers, Inc., 1994.

White, Barbara et al, eds., *The Self-Help Sourcebook: Finding and Forming Mutual Aid Self-Help Groups*. 4th ed. Denville, NJ: St. Clares-Riverside Medical Center, 1992.

White, Burton. *The First Three Years of Life*. Englewood Cliffs, NJ: Prentice–Hall, 1975.

White, William. *Slaying the Dragon: The History of Addiction, Treatment and Recovery in America*. Bloomington, IL: Chestnut Health Systems/Lighthouse Institute, 1998.

Wilson, Bill. *Bill W.: My First 40 Years*. Center City, MN: Hazelden, 2000.

Wing, Nell. *Grateful to Have Been There*. Park Ridge, IL: Parkside Publishing Corp., 1992.

Wiseman, Jacqueline. *The Other Half: Wives of Alcoholics and Their Social-Psychological Situation*. New York: Aldine de Gruyter, 1991.

Woititz, Janet. *Adult Children of Alcoholics*. Deerfield Beach, FL: Health Communications, Inc., 1980.

Wood, Barbara. *Children of Alcoholism: The Struggle for Self and Intimacy in Adult Life*. New York: New York University Press, 1987.

Woodcock, George et al, *From Prince to Rebel*. Montreal: Black Rose Books, 1990.

Woodman, Marion. *Addiction to Perfection: The Still Unravished Bride*. Toronto: Inner City Books, 1982.

Wright, Ronald. *Stolen Continents: The "New World" through Indian Eyes since 1492*. New York: Viking, 1992.

———. *A Short History of Progress*. Toronto: House of Anansi Press, 2004.

Wuthnow, Robert. *"I Come Away Stronger": How Small Groups Are Shaping American Religion*. Grand Rapids, MI: W. B. Eerdmans, 1994.

Name Index

Abrams, Nancy E. 8, 42, 57, 58, 102, 172, 186, 188, 206, 231, 267
Al-Anon Family Groups (Al-Anon) xv, 2-9, 11, 15, 19, 24, 26-28, 69, 70, 74, 101, 102, 104, 105, 108, 110-13, 115, 119, 124, 135, 150, 153, 160-73, 175, 177-81, 184, 185, 194, 195, 199, 201, 203, 206, 214, 224, 231, 238, 239, 247, 250, 271, 273, 276, 284, 286, 288, 297, 298
Alateen 5, 112, 161, 163, 169, 206, 271, 276
Alcoholics Anonymous (A.A.) xiii, xv, xix, xxi, 1-9, 11, 13-32, 38, 42, 43, 73, 80, 101, 103, 110, 116, 119, 147-55, 157, 158, 159, 162, 173-77, 188, 197, 202, 207, 216, 228, 231, 233, 235-38, 240-49, 272, 274, 275, 276, 284, 289, 295, 297
A.A. xiii, xv, xxi, 1-19, 21-32, 40-42, 45, 70, 73, 74, 91, 101, 102, 105, 109-13, 118, 119, 122, 146-52, 154-64, 167, 172-80, 185, 186, 189, 192, 195, 202-04, 206-08, 214, 215, 218, 226, 231, 232, 235, 237-39, 241-51, 271, 273-76, 281, 284, 287, 288

Alcoholics Anonymous (the Big Book) 2, 3, 5, 15, 17, 18, 22, 23, 25, 26, 103, 104, 149, 150-52, 188, 231, 237, 241, 245, 249, 272, 274
Alexander, Bruce 23, 46, 51, 58
Alexander, Jack 5
Allen, Paula Gunn 75, 79, 86, 90, 92, 98, 101, 205
Amnesty International (Canada) 119, 261
Andrews, Ted 115
Armstrong, Karen 68, 69, 142
Austin, Gregory 272
Avrich, Paul 32, 34, 38, 42
Aztec 78, 79

Barclay, Harold 13
Bateson, Gregory 122, 232
Beagle 54, 61, 77, 87, 90, 97, 99, 256, 257
Benson, E. F. 82
Benyus, Janine 82, 137, 144, 191, 264-66
Bookchin, Murray 66, 74, 79, 188
Borkman, T. 12
Bornstein, David 42
Boucher, Douglas 47, 48, 54
Bowlby, John 62, 63
Bowles, Samuel 58
Brand, Russell 8, 233
Brandt, Willy 188

Breton, Denise 7
Bridges, Lucas E. 85-88, 96, 99, 255, 268, 277
Brock, Thomas 265
Brown, Lester 188, 269, 282
Brown, Stephanie 112
Browne, Janet 55, 61, 62, 63, 82, 97, 98, 390
Browne-Miller, Angela 5, 161, 175, 219
Brundtland, Gro Harlem 188
Buchman, Frank 16
Bufe, Charles 3, 4, 32
Butler, Katy 206
Button, Jemmy 82, 83, 84, 270, 277

Campbell, Joseph 119, 321, 232
Campfens, Hubert 205
Cann, R.L. 259
Capra, Fritjof 67, 188, 209, 212
Carroll, Joseph 67
Carson, Rachel 188
Cathcart, Thomas 281
Cavanagh, John 171, 210, 212, 254
Cheever, Susan 1, 155, 233, 275
Cherokee 78, 79, 80, 90
Chesnut, Glenn F. 8
Chopra, Deepak 7
Clancy, Kelly 59
Coelho, Paulo 272, 276, 280, 294
Controllers Anonymous 196
Coyhis, Don 110, 254

Darwin, Charles iii, xv, 9, 31, 34, 44-47, 73, 77, 82-84, 87-90, 92, 95-99, 124, 125, 127, 129, 223, 226, 243, 253, 255-57, 268, 277, 279, 290
Dashuk, James 255
Dass, Ram 145
Dennett, Daniel 68
De Paolo, Charles 60, 84, 253, 255, 256, 277
Desmond, Adrien 53, 55, 61-63, 97, 98
Diamond, Jared 127, 128, 188
D'Orbigny, Alcide 92
Donne, John 126
Dossey, Larry 136
Doyle, Robert 2
Drake, Francis 77, 81, 82, 84, 95-97
Dubos, Rene 57, 188, 259, 279, 284, 285, 287, 290, 291, 293, 295
Dupont, Robert 6

Ehrlich, Paul 205
Ehrman, Max 217, 277
Elk, Black 68, 87, 220

Feest, Christian 93
Ferguson, Marilyn 190, 191
FitzRoy, Robert 63, 77, 82, 92, 97, 255
Fossey, Dian 49
Fuller, Robert 7

On Evolution

Gelman, Joseph 174
George-Kanentiio, Doug 92, 93
Gilliam, Marianne 231
Good Tracks, J.G. 94
Graveline, F.J. 74, 121
Gray, Lesley 256
Greg, Richard 172
Gusinde, Martin 84-88, 96, 255, 268

Haeckl, Ernst 256
Hager, Lori 159
Hammer, Armand 281
Hammer, Leon 248
Hammond III, William 7, 116, 186, 200, 206, 212
Hampden, John 84, 96
Hardin, Garret 72, 122
Harman, Jay 266
Harries-Jones, Peter 232
Hartigan, Francis 1, 2, 19, 238
Hawken, Paul 135, 208, 209, 212, 213, 224
Hazlewood, Nick 82, 83, 98, 99
Hazzard, Rowland (Rowland H.) 9, 15, 16, 17, 22
Heyman, Gene M. 7
Himmelfarb, Gertrude 55, 62, 63
Hirschfield, Jerry 6
Hobbes, Thomas 95
Hornbacher, Marya 2, 8, 102, 107, 158, 160, 287, 290
Horse, Crazy 68, 77, 92, 93, 112, 220

Humphreys, Keith 12
Huxley, Aldous 232

Inca 78, 79
Iroquois 75, 78, 79, 80, 86, 92, 93, 254

Jacobson, Bertil 159
James, William 9, 11, 19, 23, 70, 152, 216, 235, 248, 249
Jastrow, Robert 63, 125, 256
Joyner, Tim 95
Jung, Carl 9, 11, 15, 16, 17, 22, 28, 38, 70, 76, 134, 138, 216, 218, 216, 218, 232, 235, 262-64, 282

Kaminer, Wendy 12
Kasl, Charlotte 223, 231, 255, 273
Katz, Alfred 12-14, 23, 47, 70, 226
Kawanabe, Hiroya 66
Keating, Thomas 1, 23, 39, 158, 176, 188, 191, 221
Kelsey, H. 95
Khanna, Parag 188, 222
Kingwell, Mark 136, 137, 140-44
Kitchell, James 265
Korey, Kenneth 256
Kropotkin, Peter 9, 11, 14, 26, 28-47, 49, 51, 54-57, 63, 65-67, 69, 70, 73-75, 81, 83, 84, 88, 89, 96, 125, 129, 130,

190, 215, 216, 218, 222-26, 235, 236, 238-42, 245-51, 255
Kulchyski, Peter 21, 75, 78, 81, 87, 111, 114
Kurtz, Ernest 13, 27, 31, 32, 192, 202, 235

Lachance, Albert 197
Lavoie, Francine 12
Lawson, Anne 3
Lee, Dora 267
Leershen, Charles 206
Lestringant, Frank 92, 93
Lofland, John 9
Lovelock, James 136, 188, 199, 282
Loye, David 67

Magellan, Ferdinand 77, 81, 94-97
Makela, Klaus 3, 4, 6, 7, 12, 13, 24, 110, 148, 158
Mander, Jerry 171, 254-56, 261, 278, 282, 283
Maracle, Brian 109
Marean, Curtis 58, 59, 140, 254
Margulis, Lynn 47, 66
Marshall III, Joseph 93, 112
Massie, Joseph 198, 199
Matthiessen, Peter 77
Maya 78, 79
McCalman, Iain 98
McCarthy, R.G. 5
McGovern, John 7

McGovern, Patrick 273
McKew Parr, C. 81, 82
McKinley, Daniel 88
Mead, Margaret 1
Means, Russell 91
Menaker, Esther 159
Mencken, H.L 137, 267
Miller, William 37
Mitroff, Ian 199
Montagu, Ashley 56, 62
Montaigne, Michel 136
Morris, Brian 46, 54-56, 66, 240
Morris, Gwen 255
Mowrer, O.H. 6
Mumford, Lewis 57
Murphy, Brian 127-29, 134, 135, 138, 142, 145, 189

Nabokov, Peter 75, 76, 80, 103, 220
Neuman, W.L. 9
Newburg, Andrew 267
Novak, Michael 205
Nowak, Martin 49, 54, 57-59
Nowinski, Joseph 8, 12, 57, 102, 186, 195, 206, 207, 231, 284

O'Murchu, Diarmuid 128, 129, 143, 145

Palmer, Trevor 67
Paton, Alan 94, 183, 228
Patton, M.Q. 9

On Evolution

Peale, Norman Vincent 5
Peck, M. Scott 2, 6, 112, 175, 194, 205, 230
Peckham, Morse 64
Pittman, David 6, 25
Powell, Thomas 12

Quammen, David 63

Rank, Otto 159
Raphael, Matthew 1, 3, 4, 13, 18, 19, 21, 30, 174, 203, 233, 249
Ridley, Matt 30, 58, 59
Rifkin, Jeremy 188
Roberts, Monty 115
Robertson, Nan 6, 158
Robinson, David 5, 37
Romeder, Jean-Marie 12
Room, Robin 7, 107
Rousseau, Jean-Jacques 269, 277
Ryley, Nancy 25, 123, 138, 205, 263, 265, 268, 270, 278

Sabini, Meredith 76, 134, 138, 263, 264, 282
Saloman, Frank 55
Sandoz, Mari 68, 92
Sapp, Jan 66
Saul, John Ralston 129
Schaef, Anne Wilson 6, 105, 193
Schumacher, E. F 188
Seiberling, Henrietta 19, 20, 22
Serrano, Miguel 232

Shepard, Paul 188
Simos, Miriam 7, 68, 76, 180, 278
Sioui, Georges E. xv, 74, 86, 91, 92, 100, 187, 202, 203, 278
Smith, Bob 111, 238
Smith, Anne (Anne S.) 5, 9, 14, 19, 20, 22, 111, 218, 231, 248
Smith, Robert (Dr. Bob) 1, 3, 5, 9, 14, 16, 18-23, 111, 203, 214, 218, 231, 236-38, 248, 249, 274, 275, 281
Speerstra, Karen 59, 277, 278
Spencer, Herbert 31, 43, 231
Spretnak, Charlene 188
Standage, Tom 253, 272
Standing Bear, Luther 193
Stoneking, M. 259
Stringer, Christopher 259
Sullivan, Dennis 13
Suzuki, David 135, 142, 188, 199, 206, 208, 227

Thacher, Ebby (Ebby T.) 9, 16-18, 20-22, 103, 150, 152, 236, 237, 249
Thayer, Bradley 67
Thomsen, Robert 2, 238
Thomson, Keith 125
Todes, Daniel P. 46, 54
Tolle, Eckhart 282, 286-89, 292, 294
Tolstoy, Leo 224, 229
Tomasello, Michael 58

Torrance, Tom 89
Toynbee, Arnold 232
Trager, William 66
Trevino, A.J 119
Tze, Lao 121

UNICEF (United Nations International Children's Emergency Fund) 10

Valverde, Mariane 7
Van der Dennen, Johan 67

Wackernagel, Mathis 76
Wallace, A.R. 46, 98
Wallis, Claudia 260
Walsch, Neale Donald 191
Ward, Barbara 188
Wasserman, Harry 12
Weatherford, Jack 75, 188, 229, 230

White, William 4, 13, 16, 30, 110, 158
Wilson, A.C. 259
Wilson, Bill (Bill W.) xiii, 1-5, 11, 13-32, 36, 38, 41, 43, 44, 73, 80, 101, 102, 107, 143, 147-61, 171, 174, 176, 177, 188, 196, 197, 202, 203, 208, 214-16, 218, 228, 230-33, 235-51, 271, 275, 276, 281, 284, 289, 295
Wilson, Lois (Lois W.) 4, 9, 14, 19, 22, 25, 28, 143, 161, 169, 203, 218, 231, 238, 239, 247-49
Wing, Nell 21, 156, 158
Wood, Barbara 105
Woodcock, George 33-35, 39
Woodman, Marion 6, 25, 105, 123
Wright, Ronald 37, 59, 78, 79, 90, 119, 134, 135, 255, 259, 277

Subject Index

Ambiguity 130-34, 136, 137
Ambivalence 68, 132, 133
Anonymity 170, 203, 241, 243, 244, 274, 275
Authority xxi, 13, 26-29, 38-44, 70, 77-81, 88-90, 92, 103, 116, 118, 122, 190, 196, 202, 214-16, 218, 222, 223, 225, 241, 250, 276, 284

Benign Anarchy xvii, 28-32, 73, 80, 99, 103, 107, 119, 190, 201, 214, 223, 225, 235, 254, 276, 284
Bipolar Disorder 158
Blue Green Politics 10, 192, 193, 196-98, 200, 204, 207, 208, 214

Chewing on Ideas 123
Citizen 1, 5, 136, 42, 141, 142, 145, 147, 154-57, 177, 185, 187, 189, 190, 197, 198, 200, 201, 205, 207-09, 212, 213, 217, 249
Citizenship 143, 144, 160, 163
Competition 9, 25, 26, 38, 41, 45-49, 51-59, 61, 62, 65, 67-69, 73, 76, 89, 124-26, 129, 140, 186, 208, 223, 226, 227, 243, 250, 280

Cooperation 9, 25, 30, 46, 48-54, 56-59, 62, 64, 67-69, 72, 73, 76, 125, 129, 140, 210, 220, 279, 280
Creation 1, 53, 73, 114, 115, 119, 121, 122, 126, 127, 129, 130, 135, 137, 139, 140, 144, 148, 171, 182, 184, 186, 189, 191-94, 198, 201, 208, 212-14, 217, 218, 223, 227, 260, 262, 266, 269, 270, 279
Creator 21, 64, 65, 77, 82, 114, 116

Darwinist 37, 41, 43, 46, 53, 127, 228, 235, 239, 250, 274, 285, 291
Darwinism 30, 31, 55, 66
Dictatorship 29, 55, 250

Ecological 4, 8, 10, 14, 69, 71-78, 94, 95, 98, 100, 101, 114, 119, 120, 125, 126, 128, 130, 131, 133, 134, 136-43, 146, 172, 173, 180, 183, 186, 187, 189, 190, 192, 193, 195, 197, 200, 201, 204, 205, 207-209, 212, 213, 217-19, 221, 223, 225-27, 235, 253, 255-57, 262-65, 269, 276, 278, 279, 280, 281, 282, 285, 291,

anti-ecological 75, 94, 138, 143, 189, 208, 225, 282
Ecologically 120, 211, 213, 267, 281, 282, 285, 291
Ecology 57, 67, 74, 114, 135, 204, 256
Environment 24, 79, 123, 130, 153, 171, 199, 205, 207, 208, 212, 213
Environmental 25, 36, 72, 76, 78, 120, 135, 165, 180, 197, 198, 205, 208, 211, 216, 227, 283
Environmentalism 41, 207

First Nations iii, xv, xv, 9, 10, 13, 34, 40, 59, 61, 62, 70-82, 84-87, 91-96, 100, 101, 103, 104, 106, 107, 109-20, 121, 127, 128-130, 140, 142, 144-47, 173, 178, 183, 185, 190, 192, 195, 198, 201-03, 214, 215, 218, 219, 220-22, 225, 227, 228, 231, 232, 235, 250, 253-56, 262, 264, 268, 270, 273, 277, 283-85, 287, 291, 293
First Nations circles xv, xvii, 9, 13, 100, 101, 103-06, 108, 109, 115-19, 142, 143, 178-80, 201-04, 217, 218
First Nations societies 14, 27, 38, 40, 70, 74-78, 80-82, 86, 91-93, 100, 103, 105, 110, 114, 117, 197, 217, 218, 220, 232, 292, 345
Flying Blind 22, 23

Galapagos Islands 63, 256, 257
God xix, xxi, 8, 18, 20, 26, 39, 40, 65, 68, 69, 89, 115, 118, 121, 133, 136, 155, 156, 164, 171, 172, 176, 178, 179, 181, 182, 183, 184, 190, 191, 193, 196, 198, 202, 214, 220, 222, 223, 226, 227, 229, 249, 253, 261, 262, 264, 268, 271-73, 277, 280, 284-87, 290, 292
Goddess 259, 269, 270, 279
God-given 196, 222, 280, 287
Godless 39, 126

Higher Power 9, 11, 13, 14, 18, 20, 26, 27, 38, 40-42, 44, 65, 70, 77, 79, 80, 102, 103, 110, 111, 113, 116, 118, 123, 140, 149, 170, 176, 179, 182, 184, 187, 190, 196, 198, 215, 218, 222, 223, 225-27, 237, 250, 268, 270, 273, 290
Higher Powered Mutual Aid (HPMA) 9, 10, 14, 15, 24, 26-28, 30, 42, 73-76, 81, 83, 85, 86, 94, 100, 101, 103, 105, 118-20, 130, 140, 141, 143, 147, 149, 150, 153, 162, 164, 173, 174, 176, 177, 181, 184, 186-88, 190, 192, 194,

195, 197, 202, 204, 206, 211, 214-23, 227, 232, 247, 251, 263, 270, 273, 279, 280, 284, 285, 291, 293
Humanly knowable truth 10, 131, 136, 137, 140, 176, 218, 289
Humility 16, 41, 89, 100, 116, 156, 190-92, 220, 241, 244, 265, 268, 271, 272, 280, 290, 293, 295

Idle No More 118
Imperfection 140, 141, 146, 192
Individual in relation 10, 27, 43, 121, 122, 126-30, 132-34, 137, 138, 140, 143-46, 176-80, 182, 184, 189, 190, 192, 193, 195-98, 201, 204, 207, 214, 219, 223, 225, 242, 245, 253, 262, 265, 280, 285, 289, 292, 293
Ipperwash 118

Kropotkinism 57, 67
Kropotkinist 30, 38, 41, 66, 69, 70, 225

Leadership 3, 33, 70, 77, 86, 88, 93, 108, 109, 115, 172, 173, 177, 214, 219

Mutual aid 9, 11-15, 20, 23, 25-32, 36-41, 43, 45-54, 56-58, 61, 64, 68-70, 75, 83, 89, 104, 125, 126, 129, 140, 145, 187, 205, 215, 218, 219, 222, 223, 225-27, 236-39, 241, 246, 247, 250
Mutualism 47, 54, 66, 69

Natural Selection 45, 46, 48, 50-57, 59, 61, 64, 66-69, 89, 125, 257

Oka 118
Oxford Group 16-19, 21-23, 38, 49, 152, 237, 248

Quietism 174, 203

Racism 60, 171, 215, 220, 289
Religion 11, 17, 20, 25, 28, 38-41, 43, 46, 114, 152, 165, 167, 193, 197, 215, 221, 231, 232, 249-51, 267, 270, 283
Religious 3, 7, 14, 16, 18, 19, 22, 23, 40, 43, 102, 200, 205, 221
Respect 37, 87, 106-10, 114, 140, 141, 148, 152, 180-84, 187, 188, 201, 219, 250, 268

Sanity xvii, xlx, 23, 24, 26, 107, 109, 142, 144, 146, 151, 162, 175-77, 180, 185, 187, 194, 197, 206, 220, 244, 273, 295

Serenity xvii, 23, 24, 26, 70, 77, 80-85, 94, 99, 100, 107, 109, 111, 116, 119, 131, 141, 142, 144, 146, 151, 152, 155, 162, 167, 175-77, 180, 181, 184-87, 190, 194, 197, 206, 214, 220, 244, 272, 273, 275, 295, 274-77, 279, 295
Sleepers of Safety 104, 105, 142
Sobriety xvii, 3, 17, 21, 23, 24, 26, 107, 109, 113, 142, 144, 146, 151, 152, 162, 165, 174-77, 180, 185, 187, 194, 197, 206, 220, 247, 273, 275, 295
Spiritual Awakening xix, 11, 18-20, 38, 102, 110, 115, 149, 152, 153, 162, 164, 166, 247

Spiritual xv, xvii, xix, xxi, 1, 4-7, 10, 11, 14, 16, 18-21, 27, 30, 38-40, 71, 73, 76, 100, 101-04, 109-11, 114, 116, 118, 119, 122, 123, 129, 146, 149, 150, 152, 153, 161, 162, 164, 169, 181, 183, 187, 189, 190, 192, 193, 207, 209, 217, 218, 221, 231, 232, 235, 237, 248, 262, 269, 283, 285, 294, 295
Spiritualism 248
Spirituality 23, 25, 26, 28, 38, 70, 77, 78, 89, 109, 116, 158, 171, 206, 207, 215, 220, 236, 237, 247, 270, 273
Spiritually 27, 30, 207, 211, 264, 276, 285

Twelve Step fellowships 4, 14, 24, 70, 77, 92, 217, 238, 241
Twelve Step groups xvii, 4, 9, 12, 13, 23, 100, 101, 103-06, 108, 109, 111, 114-17, 142, 143, 177, 201, 204, 214, 217, 218
Twelve Step societies iii, xvii, 2, 4, 7, 11-13, 23, 24, 38, 40, 70, 72, 74, 77, 83, 103, 112, 113, 117, 118, 121, 122, 128, 145, 174, 203, 206, 207, 215, 218, 219, 221, 222, 228, 235, 250, 274, 285, 291, 292

Vision Quest 113

About the Author

The second of fourteen children, James G. Duncan was first raised in a Saskatchewan prairie town, and later on a farm in eastern Ontario. While farming, he became keenly interested in environmentalism and the welfare of humanity. He holds an MA in geography, and an MSW with a focus on community development.

Duncan has worked at the Soil Research Institute of Agriculture Canada in Ottawa, Ontario, participating in research regarding the U.S.-Canada Great Lakes Water Quality Agreement. As well, he worked at the Lands Directorate of Environment Canada in Gatineau, Quebec, first on contract, and later when he researched and wrote about the impact of city dwellers who had moved to surrounding rural areas.

Subsequently, he spent two fruitful years in Botswana as a District Officer (Lands), a Cuso International (the former Canadian University Service Overseas) volunteer. Botswana is dominated by the Kalahari Desert, and Duncan was primarily responsible for the development of plans for arable agriculture at Pandamatenga, as well as for quasi-urban improvements in the Kasane-Kazungula Planning Area.

Upon returning to his native land, Duncan was employed at the national offices of Cuso International and OXFAM Canada, both in Ottawa. At OXFAM, he arranged a non-governmental organization response to the Ethiopian famine of the late 1980s.

Later, at the Canadian Mental Health Association (Ottawa branch), Duncan was involved in community development for consumers of mental health services, and published a related report. Subsequently, he was engaged by the Pinecrest-Queensway Community Health Centre in an alcohol and drug harm reduction project.

During his employment, Mr. Duncan joined Alcoholics Anonymous (A.A.), the Al-Anon Family Groups (Al-Anon), and First Nations societies

James G. Duncan

in Ottawa. It was here that he formulated the concept of Higher Powered Mutual Aid (HPMA), undercutting Charles Darwin's insistence on competition in evolution. Duncan also published articles with Al-Anon for its literature development, and with the Ottawa Independent Writers group.

Now retired, Mr. Duncan continues his research and writing, hoping to do a part in lifting humanity onto the road to salvation. He lives with his family in Ottawa, and is currently working on a thirty-story account of his African adventures and experiences. James G. Duncan is a pseudonym.

www.ingramcontent.com/pod-product-compliance
Lightning Source LLC
Chambersburg PA
CBHW020625220526
45464CB00001B/31